Paleomagnetism of Sedimentary Rocks

This book is dedicated to my son, Peter (1989–2008),
to my daughters, Emily and Alice, and to Anna.

Paleomagnetism of Sedimentary Rocks

Process and Interpretation

Kenneth P. Kodama

Professor, Department of Earth and Environmental Sciences
Lehigh University
Bethlehem
Pennsylvania, USA

WILEY-BLACKWELL

A John Wiley & Sons, Ltd., Publication

Blackwell Publishing was acquired by John Wiley & Sons in February 2007. Blackwell's publishing program has been merged with Wiley's global Scientific, Technical and Medical business to form Wiley-Blackwell.

Registered office: John Wiley & Sons, Ltd, The Atrium, Southern Gate, Chichester, West Sussex, PO19 8SQ, UK

Editorial offices: 9600 Garsington Road, Oxford, OX4 2DQ, UK
The Atrium, Southern Gate, Chichester, West Sussex, PO19 8SQ, UK
111 River Street, Hoboken, NJ 07030-5774, USA

For details of our global editorial offices, for customer services and for information about how to apply for permission to reuse the copyright material in this book please see our website at www.wiley.com/wiley-blackwell.

Library of Congress Cataloging-in-Publication Data
Kodama, Kenneth P.
  Paleomagnetism of sediments and sedimentary rocks : process and interpretation / Kenneth P. Kodama.
     p. cm.
  Includes bibliographical references and index.
  ISBN 978-1-4443-3502-6 (cloth)
  1. Paleomagnetism.   2. Sediments (Geology)   I. Title.
  QE501.4.P35K64 2013
  552'.501538727–dc23
                                                2012011834

A catalogue record for this book is available from the British Library.

Front cover image: "Red Rocks" – original oil painting by Anna Kodama

Cover Design by: Steve Thompson

Set in 9/11 pt Photina by Toppan Best-set Premedia Limited

Printed in Malaysia by Ho Printing (M) Sdn Bhd

1   2012

Contents

COMPANION WEBSITE:

This book has a companion website:
www.wiley.com/go/kodama/paleomagnetism
with Figures and Tables from the book

The Paleomagnetism of Sediments and Sedimentary Rocks: Importance and Reliability

THE IMPORTANCE OF SEDIMENTARY PALEOMAGNETISM

Sediments and sedimentary rocks, both lacustrine and marine, are important targets for paleomagnetists who want to answer questions about global and regional tectonics, about paleoclimate, and about the behavior and history of the Earth's magnetic field. Of the 9259 results reported in the Global Paleomagnetic Database in 2009, 4971 (54% of them) were attributed to sedimentary rocks. Although estimates indicate that sedimentary rocks make up only about 8% of the total volume of the Earth's crust (Buchner & Grapes 2011), they are ubiquitous in the thin veneer of crustal rock available to geologists.

There are two significant reasons why sedimentary rock is an important target for paleomagnetic studies. The first is that sedimentary rocks give a nearly continuous record of the geomagnetic field. This is critically important to paleomagnetic studies because it allows undisputed time averaging of the secular variation of the geomagnetic field and the application of the geomagnetic axial dipole (GAD) hypothesis. The GAD hypothesis, that the Earth's magnetic field has been a dipole at the center of the Earth oriented parallel to the Earth's rotation axis, is central to the widespread and successful use of paleomagnetism in the Earth sciences. Without it, paleomagnetism would probably not be a subdiscipline of geology. Using the GAD hypothesis, paleomagnetists can calculate the paleolatitudes of rocks and the amount of vertical axis rotation that may have occurred in an area. The amount of time needed to adequately average the effects of secular variation is not easily determined. In fact, some paleomagnetists would argue that the departure of the time-averaged field from the GAD field is caused by a bias in secular variation that persists for millions of years, particularly back in the Paleozoic or Precambrian. Based on the behavior of the Earth's field over the past 5 million years, when plate motions would not be large enough to affect the observation of geomagnetic secular variation, averaging over several thousand years is generally considered a long enough time to ensure that a GAD field is observed.

The paleomagnetism of igneous rocks is much stronger than that of sedimentary rocks, so it is more robust and withstands the effects of remagnetization more easily than that of sedimentary rock paleomagnetism. However, just averaging the magnetizations of a pile of lava flows is no guarantee that enough time has passed to adequately average the effects of geomagnetic secular variation. For instance, data from the Hawaiian Volcano Observatory (Kauahikaua *et al.* 1998) show that the number of flows erupted in Hawaii per thousand years over the past 12,000 years varies from 1 to 11 in any given 1000 year period. Based on the sequence of flows erupted over the past 12,000 years and assuming that about 3000 years is needed to adequately average secular variation and obtain the GAD field, anywhere from 6 to 17 flows should be measured for a paleomagnetic study that can be reliably used to reconstruct paleolatitude or for other tectonic applications. Since the volcanic history in any particular region isn't known in detail, most workers use the amplitude of the circular standard deviation of virtual geomagnetic poles (VGPs) derived from lava flows to estimate whether secular variation has been adequately averaged. The behavior of the geomagnetic field over the past 5 million years is the only guide to the amount of secular variation, i.e. the amplitude of the circular standard deviation, expected if a sequence of igneous rocks has faithfully recorded secular variation.

Although there can be unrecognized unconformities and hiatuses in any sequence of sedimentary rocks, collecting samples from a thick stratigraphic section gives confidence that enough time has been sampled to average paleosecular variation. Knowing the average sediment accumulation rate from magnetostratigraphy, rock magnetic cyclostratigraphy, fossils, or from radiometric control can give assurance that secular variation has been averaged; even without this information, knowing the typical sedimentation rate for different lithologies can however guide sampling strategy and data interpretation (Table 1.1).

There is another reason why the continuity of sedimentary paleomagnetic records is important to paleomagnetists. Recent sediments (marine and lacustrine) and high-fidelity records from ancient sedimentary rocks allow the detailed observation of geomagnetic field behavior. A continuous record of Earth's magnetic field behavior is critical for understanding the generation of the geomagnetic field and for providing constraints on models of the geodynamo. The best constraints on geodynamo models come from records of transitional field behavior during polarity transitions (Merrill *et al.* 1996). There is a rich array of data from marine sediments showing the behavior of the field during the most recent polarity transition, the Brunhes-Matayama, some 780,000 years ago. Clement (1991) was the first to show preferred longitudinal bands of virtual geomagnetic pole paths during the Brunhes-Matuyama polarity transition using the paleomagnetism of marine sediments. The accuracy of this result was questioned and then modified by observations from igneous rocks, but it was the continuity of the sedimentary paleomagnetic record that was critical to the Clement's initial observation. Recent work on sedimentary records of secular variation of the geomagnetic field at high latitudes (Jovane *et al.* 2008) shows field behavior close to the so-called tangent cylinder (latitudes >79.1°), a cylinder parallel to the Earth's rotation axis that includes the inner core of the Earth. These workers collected 682 samples from a 16 m long marine core at 69.03°S in the Antarctic showing that the dispersion of secular variation was high (about 30°) during the past 2 myr suggesting vigorous fluid motion in the outer core. This kind of record would be nearly impossible to obtain from igneous rocks.

Finally, continuous records of geomagnetic field paleointensity variations from marine and lacustrine sediments not only allow another constraint on geodynamo models, but they can also provide a way to correlate and date marine sediments globally. The best example of this is Valet's work (Guyodo & Valet 1996, 1999; Valet *et al.* 2005) on constructing stacked rela-

Table 1.1 Typical sediment accumulation rates for sedimentary rocks

Sedimentary environment of deposition	Average sediment accumulation rate	Sampling thickness to average secular variation
Deep marine	1 cm/1000 years	~3 cm
Near-shore marine	0.1–1 m/1000 years	~0.3–3 m
Continental lacustrine	1 m/1000 years	~3 m

tive paleointensity records $S_{int}200$, $S_{int}800$ and $S_{int}2000$ over the past 200 thousand, 800 thousand and 2 million years, respectively.

The second important reason for the large number of sedimentary paleomagnetic results in the Global Paleomagnetic Database is that paleohorizontal can be unequivocally determined from sedimentary rocks. Knowing the paleohorizontal may seem trivial, but it is not always straightforward to determine for igneous rocks and it is absolutely critical for determining the paleolatitude of the rocks from the paleomagnetic vector, assuming a GAD field. Only the bedding of sedimentary rocks unambiguously gives the ancient horizontal. Intrusive igneous rocks provide no record of the paleohorizontal; it must be detected indirectly, sometimes by the rare occurrence of layered early crystallized minerals (Cawthorn 1996) or by techniques like the aluminum-in-hornblende paleobathymetric technique (Ague & Brandon 1996). One example of this approach comes from the paleomagnetic study of the Cretaceous Mt Stuart batholith. The paleomagnetism of the Mt Stuart batholith provides an important paleomagnetic data point in the argument for large-scale translation of Baja British Columbia along western North America's continental margin (Cowan *et al.* 1997). However, its anomalous paleomagnetic inclination could just as easily be explained by wholesale tilting of the batholith after it was magnetized. The small amount of tilt of the batholith, 7° according to the aluminum-in-hornblende paleobathymeter (Ague & Brandon 1996), argues for tectonic transport. It is not as convincing as paleohorizontal obtained from sedimentary bedding however, thus leaving the tectonic transport interpretation ambiguous. Another example of how the paleohorizontal of intrusive igneous rocks can only be determined indirectly comes from a paleomagnetic study of the Eocene Quottoon plutonic complex in the Coast Mountains of British Columbia. In this case, 12–40° tilting of the pluton is inferred from a regular decrease in exhumation age from west to east across the pluton determined by a transect of K–Ar ages (Butler *et al.* 2001). In the Quottoon study the amount of tilting has a large effect on the interpretation of the paleomagnetic inclination of the rocks.

Extrusive igneous rocks of course have a much better control on the paleohorizontal, either from direct measurement of the layering in the lava flows or from the bedding of sedimentary rocks intercalated between the flows, but there can still be ambiguities.

One concern to paleomagnetists is the problem of initial dip of extrusive igneous rocks, particularly of highly viscous volcanic flows such as andesitic rocks. Strato-volcano edifices can have initial, non-tectonic dips up to 35–42°. Even low-viscosity basaltic flows can have dips as large as 12° (MacDonald 1972; Francis 1993; Gudmundsson 2009). These unrecognized dips contribute error to the measurement of the paleohorizontal for extrusive igneous rocks and hence in the paleolatitude determined from these rocks. One good example of the possibility of unrecognized tilt in igneous rocks comes from the work of Kent & Smethurst (1998) in which the frequency distribution of inclinations from the Global Paleomagnetic Database are binned into different time periods (Cenozoic, Mesozoic, Paleozoic and Precambrian) to see if they are consistent with the GAD hypothesis throughout Earth's history. The Cenozoic and Mesozoic data have inclination frequency distributions consistent with the GAD, but the Paleozoic and Precambrian bins have more low inclinations than predicted by random sampling of the GAD. Even sedimentary inclination shallowing is ruled out as the cause because exclusively igneous results show the effect. While Kent and Smethurst speculate that octupolar geomagnetic fields contributed to the dipole in these distant times, Tauxe & Kent (2004) point out that uncorrected and unaccounted-for dips of igneous rocks, both extrusive and intrusive, could also explain the effect.

Sediments and sedimentary rocks therefore have an important role to play in paleomagnetic studies because of the continuous record they provide and because their bedding planes give an unequivocal record of the ancient horizontal.

ENVIRONMENTAL MAGNETIC RECORD FROM SEDIMENTS AND SEDIMENTARY ROCKS

The continuity or near continuity of the sedimentary record, particularly when compared to igneous rocks, also makes sedimentary rocks an important target for paleoenvironmental studies. In environmental magnetic studies, the magnetic minerals of sedimentary rocks can record paleoenvironmental conditions. While sedimentary paleomagnetism uses directional and intensity records of the geomagnetic field for correlation, dating, and paleolatitude information, environmental magnetism uses parameters that measure the concentration of magnetic minerals, magnetic particle grain size, and magnetic mineralogy of sedimentary rocks as proxies of the paleoenvironment. Several important books have been written on environmental magnetism (Thompson & Oldfield 1986; Evans & Heller 2003) and the reader is referred to them for a more complete treatment. Chapter 8 will however introduce and cover rock magnetic cyclostratigraphy in detail, which uses environmental magnetic principles. Rock magnetic cyclostratigraphy is an exciting new use of mineral magnetic measurements that provides high-resolution chronostratigraphy for sedimentary rock sequences.

Astronomically driven climate cycles are known to be recorded by sedimentary sequences lithologically (Hinnov 2000), but the environmental magnetics of the rocks can be a very sensitive detector of Milankovitch-scale climate variations. The rock magnetic record becomes particularly important when climate-driven lithologic changes are difficult to identify in sedimentary rocks (Latta *et al.* 2006). Ultimately, rock magnetic cyclostratigraphy can provide 20 kyr resolution, much better than even the best magnetostratigraphy that records even the shortest geomagnetic polarity chrons. Pioneering efforts by Ellwood (e.g. Ellwood *et al.* 2010, 2011) looking at magnetic susceptibility variations in stratigraphic type localties have not focused exclusively on the magnetic response to astronomically driven climate cycles. Measurements that examine the concentration variations of only depositional remanent magnetic minerals (magnetite or hematite) can be more straightforward to interpret than susceptibility measurements that respond to concentration variations of diamagnetic (calcite, quartz), paramagnetic (iron-bearing silicates), and remanent magnetic minerals. Concentration variations of remanent magnetic minerals therefore have the potential to provide cleaner records of global climate cycles, either run-off variations from the continents or global aridity.

THE EVIDENCE FOR HIGH-QUALITY PALEOMAGNETIC DATA FROM SEDIMENTARY ROCKS

A main focus of this book is the accuracy of sedimentary paleomagnetic records, including some processes that can cause inaccuracies and biases in sedimentary paleomagnetism. The starting point of this discussion must however be the understanding that sediments and sedimentary rocks can and do provide very high-quality and accurate records of the Earth's magnetic field throughout geologic time. The intent of this book is not to give the impression that there are insurmountable problems with sedimentary paleomagnetic data. Rather, the aim is to discuss some of the very important inaccuracies that can arise in sedimentary paleomagnetic data. These inaccuracies tend to be essentially second-order effects, but prevent paleomagnetism from achieving its full potential as an important tool for the Earth sciences.

The evidence that young sediments can provide a good record of the geomagnetic field is plentiful. I will show some examples, but it is by no means an exhaustive list. In doing so I will also give an estimate of the repeatability of these sedimentary records of the geomagnetic field; this is not so much a rigorous measure of the accuracy of the sedimentary paleomagnetic recorder, but a way of estimating the precision of the very best sedimentary paleomagnetic recorders. This is probably the only way to understand the accuracy of the paleomagnetism since, for most cases, the true direction and intensity of the geomagnetic field is not known. In this approach I will rely on studies in the scientific literature that report on multiple records from cores that sample sediments, both lake and marine, of the same age. The best records come from the most recent sediments which have not yet been appreciably affected by post-depositional processes, chemical and physical, that can affect the magnetization's accuracy and precision. Some of these studies report the scatter in inclination and declination downcore, others simply show plots of the agreement between multiple records of the field from which the scatter can be estimated. I do not calculate statistical parameters from these records but simply show, from digitizing the plots, the range of scatter in inclination and declination. The point of this exercise is to give the reader a better feeling for the repeatability of paleomagnetic records of the field and hence an estimate of their accuracy.

Before embarking on an examination of multiple records of the recent geomagnetic field, we will briefly consider how sediments become magnetized parallel to the Earth's field. The process by which sediments become magnetized is called the depositional or detrital remanent magnetization (DRM) process. In this process individual iron oxide, sub-micron-sized magnetic particles, typically magnetite (Fe_3O_4), become oriented so their magnetic moments are statistically biased toward the ambient Earth's magnetic field during deposition. There is one basic theory for how this process occurs in nature, and all modifications of this theory will be covered in detail in Chapter 2. Geologists typically sample finer-grained sedimentary rocks, e.g. siltstones, very fine sandstones, mudstones, and shales, because it ensures a quiet depositional environment that should enhance the accuracy of the DRM recording process. Geologists avoid (as much as possible) sampling rocks that have obviously been affected by significant alteration, early diagenesis, recent weathering, hydrothermal events, significant metamorphism, or heating by nearby igneous bodies. Organic-rich lake and coastal marine sediments can be affected significantly by sulfate reduction diagenesis, and should be studied paleomagnetically with this possibility in mind. Rusty stains on the outcrop could indicate production of secondary Fe oxide/hydroxide magnetic minerals such as limonite by surface weathering, so samples should be taken from as fresh a subsurface as possible to avoid recent secondary magnetizations. The growth of secondary magnetic minerals by hydrothermal or tectonic fluids can be quite subtle to see in outcrop. Petrographic observation of the rock combined with field tests of age of magnetization (fold tests, contact tests, conglomerate tests; e.g. McElhinny & McFadden 2000; Tauxe 2010) may therefore be needed to determine if these processes could have affected the paleomagnetism of a sedimentary rock.

As we look at recent sedimentary records of the geomagnetic field, another point to remember is that there has been natural variability in the direction and intensity of the Earth's magnetic field due to secular variation. The secular variation of the field, particularly the field over the past 5 Ma, has been the focus of intense study in order to better understand the processes that generate the field. Five million years is the cut-off because plate tectonic motions will not be large enough over this time period to contribute to the variability of the geomagnetic directions. Typically, data from igneous rocks are used since igneous rocks tend to yield strong, reliable paleomagnetic directions while giving nearly instantaneous readings of the field. The data are usually reported in terms of the angular standard deviation of the dispersion of virtual geomagnetic poles (VGPs). The angular standard deviation S is defined:

$$S^2 = \frac{1}{N-1} \sum_{i=1}^{N} \Delta_i$$

where N is the number of samples and Δ_i is the angular distance between a sample's direction and the mean direction. The angular standard deviation includes 63% of the samples about the mean. A VGP is simply the position of the north magnetic pole that would produce the observed paleomagnetic direction, assuming that the field has been a dipole at the center of the Earth. For a suite of directions that record secular variation it is a good way to compare data from different places on the Earth's surface, essentially normalizing for the differences expected for a dipolar field observed at different site latitudes. For a collection of 3719 lava flows and thin, quickly cooled dikes, the angular standard deviation for the VGPs from samples collected at different latitudes is seen to increase from about 11° at the equator to about 20° at the poles (McElhinny & McFadden 1997).

But what does this mean for the directional dispersion of geomagnetic field directions? Constable & Johnson (1999) provide statistical models of geomagnetic field secular variation recorded by McElhinny and McFadden's dataset, and show that the dispersion of VGPs is consistent with directional angular dispersion that ranges from about 17° at the equator to about 10° at the poles (opposite to the latitudinal dependence observed for VGPs). Furthermore, as Tauxe has shown in her elongation–inclination work on inclination shallowing corrections (see Chapter 5), while VGP dispersions are nearly circular for the recent geomagnetic field, directional distributions are elongated in a north–south orientation. The elongation of the elliptical directional distribution is greatest at the equator (nearly 3:1) but close to circular at high latitudes. These observations of the geomagnetic field set the context for appreciating the evidence for accurate DRMs in recent sediments and sedimentary rocks.

EVIDENCE OF ACCURATE EARLY DRMs

One example of how fine-grained marine sedimentary rocks can provide a reproducible record of the intensity variations of the Earth's field comes from the pioneering work of Guyodo & Valet (1996). This work has gone through several iterations, but in the first study Guyodo and Valet took relative paleointensity records derived from marine sediments and stacked them to remove the effects of noise in the recording process, thus obtaining a master relative paleointensity curve over the past 200 ka. Subsequently, this work has been extended to the last 800 ka and then to the past 2 Ma (Guyodo & Valet 1996, 1999; Valet *et al.* 2005). In the 200 ka study, they used 18 normalized paleointensity records from the world ocean and produced the $S_{int}200$ record of geomagnetic field intensity variations. They have normalized their stacked record to have a mean intensity equal to unity and can therefore report a running standard deviation for $S_{int}200$. The average value of the standard deviation is about 0.37 and gives a good sense of the repeatability of relative paleointensity DRM records. Some of this variability is due to non-dipole field effects because the records were collected globally and the non-dipole field is expressed locally, and some variability is due to errors in the recording process. It is not possible to separate the two effects in the $S_{int}200$ record, but Brachfeld & Banerjee's (2000) relative paleointensity work on Lake Pepin sediments offers some insight into the relative importance of non-dipole and recording error effects.

Directional accuracy of the paleomagnetism of recent sediments can be assessed in Lund & Keigwin's (1994) record of paleosecular variation of the field (PSV) from North Atlantic Ocean Bermuda Rise marine sediments (Fig. 1.1). Lund and Keigwin compare the natural remanent magnetizations (NRMs) of two adjacent gravity cores. Digitization and evenly spaced sampling of the inclination and declination records plotted by Lund and Keigwin show that the inclination records differ by a minimum of 0.04° to a maximum of 12.9°, while declination records differ by a minimum of 0.4° to a maximum of 15.2°. The median difference in either inclination or declination for these two marine cores is about 3°. Lund and Keigwin also collected a box core from the top 50 cm of marine sediment and sampled it with four subcores. These data showed

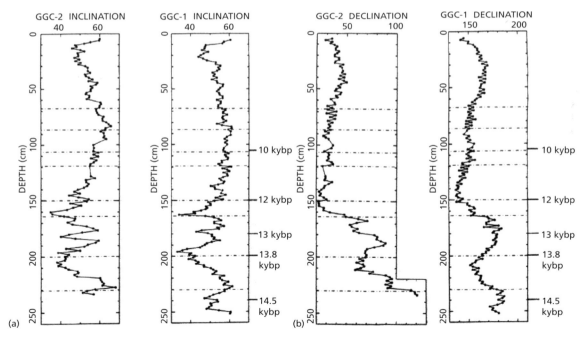

Fig. 1.1 Paleomagnetic records from the marine sediments of the Bermuda Rise showing reproducibility of paleomagnetic directions from two adjacent gravity cores: (a) inclination and (b) declination. Reprinted from *Earth & Planetary Science Letters*, volume 122, SP Lund and L Keigwin, Measurement of the degree of smoothing in sediment paleomagnetic secular variation records; an example from late Quaternary deep-sea sediments of the Bermuda Rise, western North Atlantic Ocean, 317–330, copyright (1994), with permission from Elsevier.

greater scatter in their recording of the Earth's field, probably because of the high water content of these sediments and ease with which their DRM could be disturbed by sampling. The inclinations of the subcores differed by 3.1–8.5° with a median difference of 5.6°; declinations differed by 3.6–15° with a median difference of 7.2°.

Another good example of the directional reproducibility of DRM records of paleosecular variation of the geomagnetic field comes from the data of Channell *et al.* (2004) from Ocean Drilling Program (ODP) Site 983 in the North Atlantic. In this study the authors report the Fisher mean and α_{95} confidence intervals for the averaged DRMs from multiple u-channel sampling of the core. The α_{95} of inclination varies from as little as 9° to as great as 31° with an average of 19°, while the $\alpha_{95}/\cos I$ for declination varied from 18° to 38° with a mean of 28° over the same interval in the core. Since the 95% confidence limit is twice the estimated standard error of the mean, these confidence limits for Site 983 inclinations and declinations are equivalent in magnitude to that observed for the Bermuda Rise box core.

Quite a few paleosecular variation records have been measured from both wet and dry lake sediments and all show good repeatability of DRM records of the geo-magnetic field. The multiple records of the Mono Lake geomagnetic excursion in the dry lake sediments of the Wilson Creek Formation show good reproducibility, even when the geomagnetic field is moving very quickly during the excursion (Liddicoat & Coe 1979). Before and after the excursion, the four independent records of the field differ by up to about 10° in declination and by 5–10° in inclination. During the excursion the scatter increases somewhat; one inclination record differs from the other three by 15–20° (Fig. 1.2).

More recent work on the stacked multiple paleosecular variation records from Lake El Trebol sediments in Mexico show variations at a given horizon that are about 8.5° for inclination and about 11° for declination (Irurzun *et al.* 2006). Creer & Tucholka (1982) calculate standard errors around stacked PSV curves from North American lakes and from lakes in the British Isles. Their point is to show how repeatable the DRM is in these wet lake sediments. For central North American lakes, the standard error for declination is typically 3° to as much as 23°, but averages about 8°. The standard error in the inclination record is much better, averaging only 2°. When the PSV records from Lake Windermere are compared to those from Loch Lomond the standard error is even less (about 2° for declination and 1° for inclination). Stockhausen

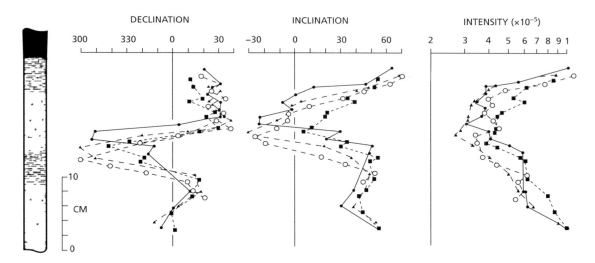

Fig. 1.2 Mono Lake geomagnetic excursion showing reproducibility of the geomagnetic field direction by sedimentary rocks, when the geomagnetic field was moving fast. JC Liddicoat and RS Coe, Mono Lake geomagnetic excursion, *Journal of Geophysical Research*, volume 84, 261–271, 1979. Copyright 1979 American Geophysical Union. Reproduced/modified by permission of American Geophysical Union.

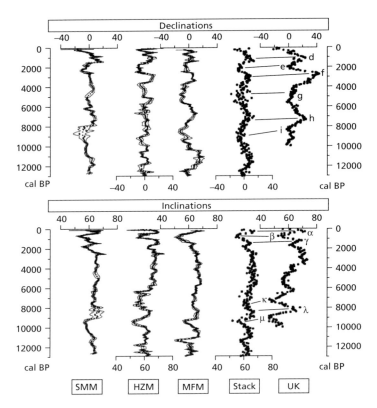

Fig. 1.3 Paleosecular variation records from the sediments of three maar lakes in Germany. The declination and inclination records of the three lakes (Schalkenmehrener Maar: SMM; Holzmaar: HZM; Meerfelder Maar: MFM) are shown with 95% confidence intervals drawn around them from multiple cores collected from each lake. The stack shows the data stacked from all three lakes, compared to the master secular variation curve from the UK. H Stockhausen, Geomagnetic paleosecular variation (0–13000 yr BP) as recorded in sediments from three maar lakes from the West Eifel (Germany), *Geophysical Journal International*, 135, 898–910, 1998, John Wiley & Sons, Ltd.

(1998) stacks the PSV records from three maar lakes in Europe and sees very good reproducibility in declination (±2.5°) and inclination (±1.5°) (Fig. 1.3). This very low scatter and high reproducibility of the field is one reason for the detailed study of maar lake paleomagnetism. Maar lakes have a very quiet depositional environment with no streams bringing in sediment loads, thus allowing the most accurate records of the geomagnetic field (Merrill *et al.* 1996).

In their study of the secular variation recorded by the DRM of wet sediments from Lake Pepin in the Mississippi River, Brachfeld & Banerjee (2000) compare the Lake Pepin secular variation record to that of nearby Lake St Croix. They report good agreement of

the inclinations with differences typically less than 5°, but in some cases they indicate that the records deviate by up to 10°. Brachfeld and Banerjee also look at anhysteretic remanent magnetization (ARM)-normalized relative paleointensity records from Lake Pepin and compare them to both relative paleointensities from Lake St Croix and variations in absolute paleointensity observed in the archeomagnetic database for the past 3000 years. The variability between these paleointensity records is reported as a standard deviation of about 0.38 around a normalized mean of unity (Brachfeld & Banerjee 2000). This is amazingly similar to that reported by Guyodo & Valet (1996) for $S_{int}200$. Since non-dipole effects are less likely to explain differences in paleointensity for these geographically proximal records, it would argue that a standard deviation of 0.38 is the best that can be expected for the reproducibility of relative paleointensity variations of the geomagnetic field; this is mainly due to recording errors.

Based on this admittedly limited, but representative, survey of the repeatability of DRM records of paleosecular variation in recent wet and dry lake sediments and wet marine sediments, DRM inclination and declination appear to be repeatable within 5–10° for either inclination or declination; the very best records show reproducibility from the same lake in the range 2–3°. Clearly this shows that recent lake and marine sediments have the capability of providing highly reproducible records of the geomagnetic field. In all these records, magnetite is the primary depositional magnetic mineral. Given the observed directional variability of the geomagnetic field over the past 5 Ma with angular standard deviations of 10–17°, the DRMs of recent sediments clearly have the resolution to discern the secular variation of the field.

DRMS IN RED, CONTINENTAL SEDIMENTARY ROCKS

The source of the paleomagnetic signal in red, continental sedimentary rocks, i.e. red beds, is quite controversial. While most geologists would agree that the magnetite carrying the paleomagnetism of recent marine and lake sediments is a primary depositional mineral, the hematite in red beds could be either depositional or a secondary magnetic mineral that grows chemically after deposition. This controversy consumed paleomagnetists in the 1970s and early 1980s and was denoted the 'red bed controversy'. Butler (1992) gives an excellent summary of the arguments for and against a DRM in red beds, as outlined during the height of the red bed controversy. In fact, red beds do carry a complicated paleomagnetic signal. Although it has been quite clearly demonstrated that the hematite in the Siwalik continental sediments in Pakistan carries a depositional remanence (Tauxe & Kent 1984), petrographic examination of some red sedimentary rocks (e.g. the Moenave Formation, Molina-Garza *et al.* 2003 or the Moenkopi Formation, Larson *et al.* 1982) of the Colorado Plateau would argue that all the hematite grains are chemically produced rather than depositional grains.

Some interesting observations about red beds add to the controversy. Most red beds sampled by paleomagnetists are Paleozoic and Mesozoic in age. Most of them are interpreted to have been deposited in fluvial environments (the molasse red beds of the Appalachians: Andreas Formation, Bloomsburg Formation, and Mauch Chunk Formation; the Colorado Plateau Paleozoic red beds: Chinle Formation, Moenave Formation), coastal marine environments (deltaic rocks: Catskill Formation), or large lake environments (Passaic and Lockatong Formations of the Newark basin), yet rocks currently forming in these depositional environments are not typically red in color. In the Global Paleomagnetic Database the youngest sediments reported to be red in color are typically of age at least 2 Ma. Therefore, the red pigmentary hematite in classic red continental sediments is most likely secondary and carries a post-depositional chemical remanence. But what about specular hematite grains that are interpreted to be detrital? As shown in the previous section, the most recent lake and marine sediments generally have magnetite as their primary magnetic mineral. There are no (or at least not many) good modern examples of detrital hematite grains being deposited in red sediments, particularly in the depositional environments usually associated with Mesozoic or Paleozoic red beds.

However, there is some very good evidence from the geologic record that continental sedimentary rocks that have micron-sized hematite grains carrying their paleomagnetic signal have been magnetized by a DRM. The presence of relatively large-grained, specular hematite is often used as evidence for a DRM (Passaic Formation; McIntosh *et al.* 1985). The best way to determine if specular hematite is present in a rock is by petrographic examination of a magnetic extract, but indirect methods such as a square-shaped thermal demagnetization intensity plot can also be used. Classic red bed units such as the Carboniferous Mauch Chunk Formation often show a very steep decrease in intensity at the highest demagnetization unblocking temperatures (Fig 1.2; Chapter 6). This behavior can be interpreted to mean that there is a very narrow grain size distribution of the hematite carrying the highest unblocking temperatures. Since hematite has a low spontaneous magnetization, its grains do not become multi-domain until relatively large grain sizes are reached (15 microns; Dunlop & Ozdemir 1997) and so all the hematite observed during thermal demagnetization of fine-grained continental sediments is typically single domain. The highest-unblocking-temperature magnetization of red beds is therefore carried by the largest single-domain hematite grains. The steep decrease in intensity during thermal demagnetization of some red beds then can be interpreted as evidence for specular hematite carrying the paleomagnetism of the red bed.

Other evidence used to argue for a DRM in red beds came during the heyday of the red bed controversy from the measurement of red bed paleomagnetism on the topset, bottomset and foreset beds of crossbedding (Steiner 1983). A difference in the inclination carried by the flat-lying bottomset or topset beds and the initially dipping foreset beds would suggest a DRM affected by an initial bedding slope by magnetic grains rolling down the slope. Another powerful argument that the magnetization of red beds is primary (and therefore a DRM) comes from the observation that many red beds record a magnetostratigraphy, in which geomagnetic field polarities are constrained to stratigraphic horizons. More recent evidence that red beds are magnetized by a DRM come from what is interpreted to be records of paleosecular variation in red sedimentary rocks. Kruiver et al. (2000) report records of paleosecular variation (PSV) in Permian-age red beds from Dome de Barrot in southeastern France. They even go as far as to analyze the characteristics of secular variation in the Permian and observe looping of the geomagnetic field vector, behavior suggesting paleosecular variation. Much earlier, Evans & Maillol (1986) measured red bed samples taken from unoriented cores and interpreted the directional variations as due to PSV. They modeled the PSV with dipole wobble plus shorter non-dipole variations, an approach typically used with the secular variation recorded by recent marine and lake sediments. Finally, depositional magnetic fabrics carried by the anisotropy of magnetic susceptibility (AMS) and magnetic remanence (AMR) of red beds suggest that their paleomagnetism is carried by depositional magnetic minerals (Kodama & Dekkers 2004). This evidence will be covered in more detail in Chapter 5.

An interesting conundrum about the accuracy of red bed remanence arises from the different possibilities for the manner in which a red continental sedimentary rock becomes magnetized. If the red bed is magnetized primarily by a post-depositional chemical remanent magnetization (CRM), then the direction carried by the red bed may be an accurate record of the geomagnetic field direction when the secondary hematite grains grew. The age of the magnetization will not however be the age of the rock, and it will be difficult to determine accurately. If on the other hand the red bed is magnetized by a DRM, then the timing of magnetization is exactly the age of the rock and will be accurately known. The direction of the magnetization will most likely not accurately record the direction of the geomagnetic field at deposition, however. The low spontaneous magnetization of hematite will cause detrital hematite grains in red beds to be more easily affected by gravity and other misorienting forces at deposition (water currents, initial slope). Tauxe & Kent (1984) showed this beautifully with their pioneering work on the Siwalik River sediments of Pakistan. Redeposition of these hematite-bearing sediments in the laboratory revealed a large inclination error as did observation of naturally redeposited sediments on the river floodplain. Tan et al. (2002) also redeposited red sedimentary rocks in the laboratory, in this case rocks from western China. They found a large compaction-caused inclination error for the finest-grained sediments and a large syn-depositional inclination error for the coarser-grained sediments. Therefore, accurate red bed directions will probably not have their age well known, and accurately dated red bed magnetization ages will probably not be an accurate record of the Earth's field direction at the time of deposition.

POST-DEPOSITIONAL PROCESSES THAT AFFECT THE MAGNETIZATION OF SEDIMENTS AND SEDIMENTARY ROCKS

Even though it's quite clear that the DRM acquired by recent sediments is reproducible when multiple coeval records are compared, this still begs the question of how accurately the DRM records the Earth's magnetic field. Chapter 2 will examine the accuracy of DRM in detail. However, even if it turns out that the DRM of recent sediments is not only reproducible but accurate, there are post-depositional processes that occur in sediments and sedimentary rocks that can either totally obliterate, distort, or bias the direction and intensity of the initial DRM acquired by the rock. These processes are: post-depositional remanence acquisition, burial compaction, reduction diagenesis, secondary growth of magnetic minerals, and tectonic strain. Not all of these processes, and in some cases none, occur in any given sediment or sedimentary rock, but we need to be cognizant of them all to be aware of their potential effects. Each process will be covered in a separate chapter in the book, along with an estimate of the magnitude of its effect on the paleomagnetism of the rock. For now, we'll give a brief overview of each.

Post-depositional remanence

An underlying assumption of post-depositional remanence, based on laboratory experimental work and field observations, is that when sediments are first deposited the pore spaces between the non-magnetic grains in the rock are larger than the submicron–micron-sized magnetic mineral particles that give the rock its magnetization. The magnetic grains can therefore remain mobile and realign their magnetic moments with Earth's magnetic field, thus becoming more accurate in their recording of the field if they became disoriented during touchdown on the sediment–water interface. As the sediment is buried and becomes dewatered by compaction the pore spaces decrease in size, trapping and immobilizing the magnetic mineral particles. Any sediment will have a grain size distribution of magnetic particles, and the largest magnetic particles in the sediment will become trapped at shallower depths below the sediment–water interface than smaller magnetic particles.

There are several results of this vision of post-depositional remanence. First, the acquisition of the sediment's magnetization, its post-depositional remanent magnetization (pDRM), will occur lower in the sediment column than where it was deposited. So even though the pDRM was acquired after the sediment was deposited, by locking in at a finite depth (typically several centimeters to 20 cm according to some work; deMenocal *et al.* 1990; Tauxe *et al.* 1996) the magnetization acquisition time will appear to precede the depositional age of the sediment. Second, it has been argued by some (see review by Verosub 1977) that pDRM is more accurate than a syn-depositional DRM because the grains have been able to reorient after touchdown to be parallel to the ambient magnetic field, and lessen the disturbance that occurs at touchdown. Finally, in this model of post-depositional remanence, different-sized magnetic mineral grains are immobilized or locked in at different depths. The result is that the paleomagnetic recording of the field is smoothed over the depth range in which all the magnetic grains are locked into place by decreasing pore size (see summary in Tauxe *et al.* 2006).

If pDRM is important in a sedimentary rock, the rock's magnetization is therefore more likely to be an accurate record of the geomagnetic field. It will however appear to occur earlier in the sediment column than its actual depositional age, so its timing will be inaccurate. It is also highly likely that the paleomagnetic record of the field will be smoothed, and the record of high-frequency changes in the geomagnetic field will be lost. The degree of smoothing will depend on the sediment accumulation rate; fast rates will minimize the effect of smoothing.

Burial compaction

At greater depths in the sediment column (from tens to hundreds of meters) than depths which would cause the lock-in of a pDRM, the dewatering due to burial compaction can cause a bias in the magnetization direction of sediments and sedimentary rocks. Experimental data from our laboratory shows that burial compaction can cause a shallowing of paleomagnetic inclination of the order 10–20°, particularly for clay-rich sediments. This effect was not fully appreciated in the early days of paleomagnetism and even during its highly productive middle age, because observations of the DRM and pDRM of recent marine and lake sediments did not typically demonstrate any bias in the recording of the geomagnetic field inclination. One of the classic experiments that informed paleomagnetists was the observation by Opdyke & Henry (1969) that cores collected at different latitudes from the world ocean recorded inclinations entirely consistent with their latitudes, following the dipole equation for the GAD:

$$\tan (\text{paleomagnetic inclination}) = 2 \times \tan (\text{latitude}).$$

The cores sampled for these measurements were only about 1–2 m in length however, so the effects of burial compaction were not evident or important for these recently deposited marine sediments.

Another reason that inclination shallowing caused by compaction was not considered important by many paleomagnetists is that it doesn't become obvious unless the sedimentary rocks were deposited at intermediate latitudes. Inclination shallowing follows a tangent–tangent relationship (King 1955):

$$\tan (\text{observed inclination}) =$$
$$f \times \tan (\text{inclination at deposition})$$

where f is the amount of shallowing between 0 and 1, with typical values between 0.45 and 0.7 for compaction-caused inclination shallowing (Bilardello & Kodama 2010b). At mid-latitudes inclination shallowing may be as much as 15–20°, whereas the effect can be 10° or less and is closer in magnitude to the resolution of typical paleomagnetic data, making the shallowing harder to detect, at high or low latitudes. This was the case for the paleomagnetic data from tectonostratigraphic terranes from western North America or central Asia. In both cases the expected paleolatitudes for the terranes, if they had not moved with respect to the craton, were intermediate in magnitude: 30–40°N for the Cretaceous Peninsular Ranges terrane and near to 45°N for the central Asian sites with anomalous paleomagnetic directions from late Mesozoic to early Cenozoic sedimentary rocks. In both cases, interpretation of the paleomagnetic data without taking account of the effects of compaction-caused inclination shallowing led to the misinterpretation that about 1000–1500 km (approximately 10–15° of paleolatitude) of northward tectonic transport had occurred.

The problem of inclination shallowing also became evident in the so-called Pangea B problem. Paleomagnetic data from the latest Paleozoic and earliest Mesozoic requires an overlap between Laurasia and Gondwana in their equatorial regions. Rochette & Vandamme (2001) have suggested that this is the result of inclination shallowing forcing both Laurasia and Gondwana closer to the equator in paleogeographic reconstructions than they actually were, thus requiring the Pangea B reconstruction in which Gondwana is shifted eastward with respect to Laurasia. A counter-clockwise 'twist' along the Tethys seaway is needed to bring the continents into their Pangea A configuration later in the Mesozoic, although there is scant geological evidence for this megashear zone (Van der Voo 1993; Torcq et al. 1997). Inclination shallowing as an explanation for the Pangea B problem would be reasonable if the size of the inclination bias were greater than the typical 95% confidence limits for the apparent polar wander path used for the paleogeographic reconstruction. In Torsvik & Van der Voo's (2002) analysis to determine the highest-quality Gondwana paleopoles at about 250 Ma, when the Pangea B overlap persists, 11 of the 12 paleopoles are from sedimentary units. Five of these sedimentary units with paleopoles closest to the mean Gondwana paleopole for 250 Ma were all deposited at intermediate latitudes in India, Pakistan,

and South America (Kamthi and Mangli beds and Panchet beds of India, Wargal and Chhidru Formation of Pakistan, and the Amana Formation of South America). In fact, Bilardello & Kodama (2010a) showed that if the Carboniferous sedimentary paleopoles for Gondwana are corrected for inclination shallowing and compared to an inclination-corrected paleopole for North America, then a Pangea A configuration is possible (Fig. 1.4).

Reduction diagenesis

In organic-rich marine and lake sediments iron-sulfur reduction diagenesis causes iron oxide minerals to dissolve and be replaced by a sequence of iron sulfides, some of which are ferromagnetic (Karlin & Levi 1983, 1985; Canfield & Berner 1987). The mineral at the end of the sequence of iron sulfides produced is pyrite (FeS_2), which doesn't carry any remanence but is paramagnetic. Both pyrrhotite ($Fe_{1-x}S$) and greigite (Fe_3S_4) are important ferromagnetic minerals formed during reduction diagenesis and will contribute a secondary chemical remanent magnetization (CRM) to the sediment. These secondary minerals are of course formed at the expense of the primary depositional magnetic mineral, magnetite, as it dissolves. Because the smallest magnetic grains have the largest surface area to volume ratio, they are preferentially destroyed in the process. The magnetic grain size is therefore seen to coarsen during reduction diagenesis and the magnetic hardness, or coercivity, of the magnetic particles of the sediment decreases.

One good example of the reduction diagenesis process comes from Lake Ely in northeastern Pennsylvania where the concentration of fine-grained biogenic magnetite is seen to decrease downcore between 30 cm and 75 cm depth in the sediment column in organic-rich lake sediments with a loss-on-ignition (LOI) of 25%. Secondary Fe sulfide magnetic phases are produced during the dissolution of the primary biogenic magnetite. The evidence is however indirect, mainly from SIRM/χ ratios (the ratio of saturation isothermal remanent magnetization, the highest magnetization acquired by a sample in a strong magnetic field, to magnetic susceptibility) that are high when Fe sulfides are present (Peters & Dekkers 2003). The depth of reduction diagenesis in marine sediments is inversely dependent on the organic carbon flux and ranges from as deep as several meters for very low fluxes of milli-

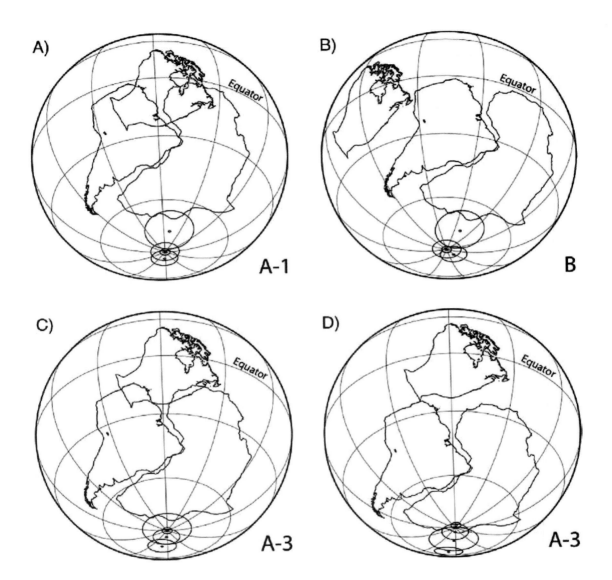

Fig. 1.4 Different Pangea configurations at 310 Ma. (a) Pangea A-1, uncorrected paleomagnetic data; (b) Pangea B, corrected North American data but uncorrected Gondwanan data; (c, d) Pangea A-3, inclination-corrected data showing the range in continental positions allowed by the paleopoles' 95% cones of confidence. Reprinted from *Earth & Planetary Science Letters*, volume 299, D Bilardello and KP Kodama, A new inclination shallowing correction of the Mauch Chunk Formation of Pennsylvania, based on high-field AIR results: Implications for the Carboniferous North American APW path and Pangea reconstructions, 218–227, copyright (2010), with permission from Elsevier.

grams of C/m^2 yr to depths of only 1 cm for tens of grams of C/m^2 yr (Schulz & Zabel 2006). The secondary ferromagnetic iron sulfide minerals that are formed essentially give a delayed NRM acquisition for the sediments affected, and the sediments are no longer magnetized by a DRM but by a CRM. Since this CRM is typically formed very early in the post-depositional history of the sediments, it could be affected by subsequent burial compaction inclination shallowing even though it will probably have an accurate directional recording of the geomagnetic field when the secondary minerals form. Efforts to use an early CRM in marine sediments for relative paleointensity measurements will be complicated by changes in paleoenvironmental conditions, causing more or less organic carbon to be deposited in the sediments (Schulz & Zabel 2006).

Later secondary CRMs

Of course, the formation of secondary magnetic minerals can occur at any stage in the post-depositional history of a rock. A rock can then have both a primary magnetization, typically carried by depositional magnetite, and a secondary magnetization, usually carried by iron sulfides or hematite. If the secondary magnetic minerals have the appropriate grain size distribution they may have higher coercivities or unblocking temperatures than the primary magnetic minerals and their remanence will be isolated by demagnetization as the characteristic remanence of the sedimentary rock. In some cases, the primary magnetization of the rock may be destroyed during the growth of the late stage magnetic minerals or the primary magnetization may be so weak initially that it is swamped by the secondary CRM. The best example of this is the nearly ubiquitous Late Paleozoic remagnetization of rocks throughout eastern North America that occurred during the Alleghanian orogeny (McCabe & Elmore 1989). This remagnetization has totally reset the remanence of Paleozoic carbonates throughout eastern North America, but it is also present as one of several components of magnetization in red clastic sedimentary rocks from the Appalachians. One possible explanation for the formation of the secondary magnetization is that fluids squeezed through the rocks of North America during the Alleghanian orogeny, causing the growth of secondary magnetite in the carbonate rocks (Oliver 1986). These fluids also apparently caused the growth of secondary fine-grained, perhaps pigmentary, hematite in clastic sedimentary red rocks.

If late stage secondary CRMs occur in sedimentary rocks, the direction of the remagnetization will most likely be an accurate record of the geomagnetic field direction. It is unlikely that a late stage CRM will be affected by burial compaction; however, the remagnetization age will be difficult to determine. Field tests such as the fold test may help constrain the age of remagnetization, but it will not be accurately known. Because the paleohorizontal cannot be determined at the time the CRM formed, particularly if the rock has been in a tectonically active area, paleopoles or paleolatitudes determined from the CRM will be dubious.

Tectonic strain

Tectonically deformed regions are important targets for paleomagnetists, primarily because folded rocks allow them to constrain the age of magnetizations in a rock using the fold test (Graham 1949). Paleomagnetic results can also provide important insights about the tectonic events that occurred in a region. Most paleomagnetists only remove the tilt of folded strata in Graham's fold test and not the effects of grain-scale strain that may have occurred during the folding. It is interesting that in his landmark paper suggesting the fold test, Graham (1949) considered that both rigid and grain-scale rotations could affect the paleomagnetic remanence. Rock deformation can indeed affect the accuracy of a sedimentary rock's paleomagnetism. Rotation of a rock's remanence can be caused by bedding-parallel simple shear strain during flexural flow/slip folding (Kodama 1988). The geometry of the simple shear not only causes an inaccuracy in the direction of remanence, but an inaccurate estimate of the age of magnetization in that the simple shear rotation can make a pre-folding magnetization appear to be syn-folding in age. This has been demonstrated in folded red beds (Stamatakos & Kodama 1991a) and shows that magnetic grains can rotate as active particles. If the remanence behaves as a passive line, then pure shear during buckle folding will rotate the remanence. Sense and magnitude of rotation will however be case specific, depending on the relative angles between the strain ellipsoid and the remanence.

Grain-scale strain in a deformed rock can cause secondary remagnetizations since secondary magnetic minerals can be formed during strain events. In some cases, the primary magnetic minerals can be destroyed when the rock is strained. Strain may preferentially affect magnetic grains of different coercivities, and the response of sedimentary rock magnetizations to strain can be quite complicated. Experimental deformation work has also shown that the domain walls of the magnetic particle can be affected by strain, thus preferentially affecting low-coercivity multi-domain magnetite grains. In other work, laboratory strain has randomized the orientation of small high-coercivity hematite grains.

When deformation causes remagnetization or scatter of directions, the effect will be easy to detect and disregard. When a subtle bias is caused however, as in flexural flow/slip folds, then independent strain measurements (in addition to paleomagnetic measurements) are needed to detect and account for the inaccuracies.

SUMMARY

In conclusion, I'd like to end with this thought: the more you know about the paleomagnetism of rocks, the more amazed you are that paleomagnetism works so well (Allan Cox, pers. comm., *c.* 1975); and it does work, quite well. Despite all the topics we'll cover in this book, the paleomagnetism of sedimentary rocks does do a very good job of recording the geomagnetic field. The goal of this book is to help geoscientists who rely on paleomagnetic data become cognizant of the possible pitfalls in understanding their data, so that the most accurate interpretations will result.

2

The Magnetization Mechanism of Sediments and Sedimentary Rocks: Depositional Remanent Magnetization

TRADITIONAL DRM THEORY: MAGNETITE-BEARING ROCKS

Even though paleomagnetists have measured and interpreted the magnetization of sedimentary rocks from the very beginning of paleomagnetism, they are still investigating and debating how sediments and sedimentary rocks acquire their paleomagnetism. This chapter will review the essentials of how sediment acquires a depositional (or, some say, detrital) remanent magnetization (DRM) and also show some of the insights that experimental compaction studies can provide for understanding the DRM of rocks. For the purposes of this discussion we will treat the magnetization of hematite-bearing sediments and magnetite-bearing sediments separately, although similar processes probably operate for both types of magnetic minerals.

There are two paleomagnetism text books that provide excellent summaries of the DRM process: Butler's (1992) *Paleomagnetism: Magnetic Domains to Geologic Terranes* and Tauxe's (2010) *Essentials of Pale-*

omagnetism. The reader is referred to these books for DRM processes presented in the context of other magnetization processes, for instance thermal remanent magnetization (TRM) for igneous rocks and chemical remanent magnetization (CRM) for secondary magnetizations in sedimentary rocks.

The traditional conception of how magnetite-bearing sedimentary rocks acquire their magnetization can be traced back to Nagata's important textbook on rock magnetism (Nagata 1961). Collinson (1965) further developed Nagata's basic idea of DRM acquisition. The main point made by these workers (the traditional view of DRM acquisition) is that the geomagnetic field causes a torque on the magnetic moment of a magnetic nano-particle, aligning it parallel to the geomagnetic field as it falls through the water column and is deposited at the sediment–water interface (see Fig. 2.1 and treatment in Butler 1992).

The classic DRM equation that mathematically represents the torque of the Earth's magnetic field on the magnetic moment of a ferromagnetic mineral grain is (Nagata 1961):

Paleomagnetism of Sedimentary Rocks: Process and Interpretation, First Edition. Kenneth P. Kodama.
© 2012 Kenneth P. Kodama. Published 2012 by Blackwell Publishing Ltd.

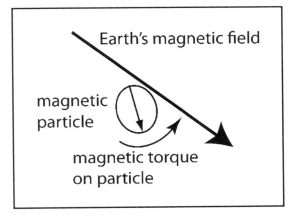

Fig. 2.1 Magnetic torque on a magnetic particle settling in the water column or settled into a large pore space after deposition. The geomagnetic field exerts a torque on the magnetic moment of the particle, both shown by arrows.

$$I\frac{d^2\theta}{dt^2} + \lambda\frac{d\theta}{dt} + mB\sin\theta = 0$$

where I is the moment of inertia of the grain (in this case assumed to be spherical, $I = \pi d^5\rho/60$); λ is the coefficient of viscous drag between the particle and water ($\lambda = \pi d^3\eta$ where η is viscosity of the water); d is diameter of the grain; and ρ is the grain density. The third term is the torque (restoring force) of the magnetic field B acting on the magnetic moment of the particle m. In traditional treatments, the small angle approximation is used and the equation becomes linear with $\sin\theta \sim \theta$. With the small angle approximation, the equation becomes that of a damped, simple harmonic oscillator meaning that the grain swings back and forth about the field direction with decaying amplitude until it becomes aligned. The second approximation made is that, for very small particles, d^5 becomes vanishingly small and the first term can be made to disappear. This gives a first-order linear differential equation:

$$\lambda\frac{d\theta}{dt} + mB\theta = 0$$

The solution of this equation shows that the alignment improves as an exponential function of time:

$$\theta(t) = \theta_0\exp\left(-\frac{t}{t_0}\right)$$

with $t_0 = \pi d^3\eta/mB$. Tauxe (2010) prefers not to use the small angle approximation and arrives at the more rigorous solution found in Nagata (1961) after neglecting the inertial term:

$$\tan\left(\frac{\theta}{2}\right) = \tan\left(\frac{\theta_0}{2}\right)\exp\left(-\frac{t}{t_0}\right)$$

For either solution, the end result is essentially the same. Because of magnetite's strong spontaneous magnetization, the magnetic grain will be more easily magnetized along its long axis (should it have a shape anisotropy) or along its 111 crystallographic axis (should the grain be equidimensional). Once the grain settles through the water column, is deposited and then incorporated into the sediment column, it should keep its alignment and provide a record of the direction of the geomagnetic field during deposition. However, as many workers have pointed out, when reasonable values for a grain's magnetic moment ($5 \times 10^{-17}\,\mathrm{A\,m^2}$), the strength of the geomagnetic field ($50\,\mu\mathrm{T}$) and the viscosity of water ($10^{-3}\,\mathrm{Pa\,s}$) are used in the traditional DRM theory equations, the time of alignment is of the order just one second.

$$t_0 = \frac{\pi d^3\eta}{mB} = \frac{3.14159\times10^{-6}\times10^{-3}}{5\times10^{-17}\times50} = 1.3$$

This result would predict that all the grains in a sedimentary rock should be perfectly aligned in fields of the same strength as the Earth's field and that DRM intensity should be independent of geomagnetic field intensity. However, early re-deposition experiments with glacial clays (Johnson et al. 1948) showed a strong linear dependence between laboratory magnetic field strength and DRM intensity. This result shows that the simple traditional notion of how a rock acquires a DRM needs modification.

Collinson (1965) tackled this problem by introducing a misaligning mechanism, Brownian motion, to mitigate the aligning torque of the geomagnetic field. He modeled the Brownian motion with Langevin theory (see detailed treatment in Butler's or Tauxe's paleomagnetism textbooks) that is also used to describe the origin of paramagnetism in materials:

$$\frac{\mathrm{pDRM}}{\mathrm{pDRM_{sat}}} = \coth\left(\frac{mB}{kT}\right) - \left(\frac{kT}{mB}\right)$$

where m is grain moment; B is the Earth's field strength; k is Boltzmann's constant; T is temperature; pDRM is post-depositional remanent magnetization; and $pDRM_{sat}$ is saturated pDRM for all the particles in the sediment perfectly aligned (Johnson *et al.* 1948). Basically, a mis-aligning force (in this case the randomizing effects of thermal energy or Brownian motion of water mole-cules, kT) counteracts the aligning torque of the geo-magnetic field acting on the magnetic moment of the magnetic particle, mB.

This approach does a good job of fitting experimen-tal results of re-deposited glacial clays (Verosub 1977) if a uniform distribution of magnetic moments with a maximum cut-off value is used (Stacey 1972). The hidden assumption of using this model is the magnetic moments of the detrital magnetic nano-particles. We used $5 \times 10^{-17}\,A\,m^2$ in our calculated estimate of the alignment time of a magnetite particle. There we assumed that the grain was one micron (μm) in diam-eter ($10^{-6}\,m$) and the magnetization J of this grain was $0.1 \times 10^3\,A/m$. If the grains are truly single domain it would be reasonable to use the spontaneous magneti-zation of magnetite ($J = 480 \times 10^3\,A/m$) or, more accurately, calculate it from micromagnetic modeling (Tauxe *et al.* 2002) ($J = (340\text{--}480) \times 10^3\,A/m$; see Tauxe 2010, fig. 4.5, p. 56). If the grains are in the pseudo-single domain or very small multi-domain grain-size range, i.e. submicron to 1–2 microns in size, then the subdivision of the grain into a small number of domains would decrease the grain moment from the spontaneous magnetization or micromag-netic modeling values; the exact magnetic moment would depend on the number of domains and their configuration.

Hence the estimate of $J = 0.1 \times 10^3\,A/m$ used by Butler (1992) and here in the alignment time calcula-tion for small multi-domain grains. A maximum grain moment of $7.4 \times 10^{-17}\,A\,m^2$ (Stacey 1972; Butler 1992, p. 72) used in the Langevin description of Brownian motion yields a curve (Fig. 2.2) that fits the re-deposition data of Johnson *et al.* (1948) and assumes something close to $J = 0.1 \times 10^3\,A/m$ for the magneti-zation of the magnetic grains ($J = 0.07 \times 10^3\,A/m$). The Brownian motion modification of classical DRM theory usually describes a post-depositional remanent magnetization in sediments and sedimentary rocks, so the alignment is envisioned to occur in the pore spaces of the sediment after the grains have been deposited. Post-depositional remanence will be covered in more detail in the next chapter.

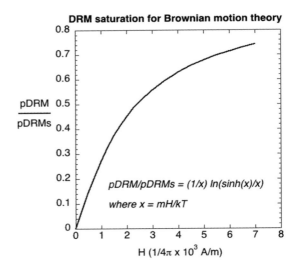

Fig. 2.2 Plot of Langevin description of Brownian motion affecting the alignment of particles in a pDRM. The curve plotted here comes from Stacey's (FD Stacey, On the role of Brownian motion in the control of detrital remanent magnetization of sediments, *Pure Applied Geophysics*, 98, 139–145, 1972, John Wiley & Sons) adaptation of the Langevin description of Brownian motion, assuming a uniform grain moment distribution with a maximum moment (m in equation for x in the figure) of $7.4 \times 10^{-17}\,A\,m^2$. Johnson *et al.*'s (1948) re-deposition data for glacial sediments fits this curve quite nicely.

FLOCCULATION MODEL OF DRM FOR MAGNETITE-BEARING ROCKS

Lisa Tauxe has provided an alternative model for DRM acquisition (Tauxe *et al.* 2006) that gets around the problem of classical DRM theory which predicts nearly instantaneous alignment of grain magnetic moments with the geomagnetic field. The main idea of her model is that the effective magnetic moment of a magnetic grain is significantly reduced because it is embedded in non-magnetic sediment floccules. The composition of the floccules is not discussed in her 2006 paper, but they could be organics, clay particles or both. Mitra & Tauxe (2009) take the flocculation idea a step further and envision clay floccules with an internal hierarchi-cal structure. They also show through a series of re-deposition experiments in solutions of varying salinity that the size of the floccules (small floccules at low salinity, large floccules at higher salinities) controls the linearity of the DRM intensity response to field strength and the accuracy of the direction of the DRM. Their theoretical modeling and experimental results would suggest that freshwater sediments can be very accurate recorders of geomagnetic field directions but not of relative paleointensities, while marine sediments will do better as relative paleointensity recorders but may suffer from inclination shallowing. The basic idea behind this observation is that the orientation of small floccules at low salinities is controlled predominately by the geomagnetic field but the DRM response to the field strength is non-linear. However, the orientation of large floccules that result from higher salinities, i.e. those in marine conditions, are controlled mainly by hydrodynamic forces. The damping effect of the hydrodynamic forces makes the response of the DRM of large floc-dominated sediments (marine sediments) more linear to the field strength, but can lead to directional inaccuracy. Moreover, small fluctuations in floc size, particularly for the small flocs of freshwater sediments, can have a large effect on DRM efficiency (another reason for inaccurate relative paleointensity variations for freshwater sediments).

The proposal that flocculation affects the DRM acquisition mechanism has its roots in Shcherbakov & Shcherbakova's (1983) theoretical treatment of magnetic and non-magnetic particles sticking together (coagulating) either during diffusion or gravitational settling. The only sticking mechanism mentioned in Shcherbakov and Shcherbakova's paper is van der Waals forces. Diffusion is predicted to be more important for particles (non-magnetic and magnetic) less than 1 μm in size, and kinematic processes are more important for larger particles (>1 μm) settling through smaller particles. Anson & Kodama (1987) independently realized the importance of clay particles and magnetite sticking to each other and proposed an electrostatic sticking mechanism. They used it to explain inclination shallowing observed in their laboratory compaction experiments. Sun & Kodama (1992) provided a more evolved model for inclination shallowing than that of Anson & Kodama, based on more sophisticated experimental evidence in which magnetite particles become attached to clay particles by electrostatic and van der Waals forces when sediment is initially deposited. As the clay fabric in a sedimentary rock develops soon after deposition, clay domains are formed and the magnetite particles become embedded in the clay domains. Clay domains have been observed in natural marine clay-rich sediments (e.g. Bennett

et al. 1981) and are essentially agglomerations or floccules of clay particles sticking together by electrostatic forces. For clay-rich marine sedimentary rocks, the floccules of Tauxe *et al.* (2006) and Mitra & Tauxe (2009) may well be the clay domains of the Sun & Kodama (1992) model. The only difference between the models is when the magnetite grains become attached to the clay grains and are incorporated into the clay domains. The Tauxe flocculation model suggests that this happens before deposition during settling. This scenario is proposed initially in Shcherbakov & Shcherbakova's (1983) theoretical paper. Sun & Kodama (1992) did not specify the timing but implied that attachment happened at or soon after deposition; this was probably because they visualized the sedimentary particles being so widely dispersed in the water column that they were essentially non-interacting until touchdown.

The syn- or post-depositional attachment of magnetite particles to clay grains would undoubtedly inhibit post-depositional realignment. The observation that the magnetization does not decrease during critical point drying of a recently deposited high-water-content sediment (Sun & Kodama 1992) would suggest that the magnetic grains are firmly fixed to the clay particles or embedded in clay domains, even in a very-high-porosity sediment soon after deposition. The lack of post-depositional realignment of magnetic particles was also an important implication of the model of Tauxe *et al.* (2006). The early laboratory re-deposition experiment by Johnson *et al.* (1948) that showed a linear dependence between field and DRM intensity was conducted with glacial clays and fresh water, so it is unlikely that flocculation was important. Electrostatic or van der Waals attachment of magnetite particles to unflocculated clay particles, as envisioned by Sun & Kodama (1992), was the probable misalignment mechanism.

DRM IN HEMATITE-BEARING ROCKS

The magnetization mechanism of hematite-bearing sedimentary rocks, in particular red beds, has long been controversial (see discussion of the red bed controversy in Butler 1992). Both a DRM and a chemical remanence (CRM) have been proposed for the paleomagnetism of red beds with a variety of evidence supporting each idea. Some workers have shown different inclinations in the topset or bottomset beds when

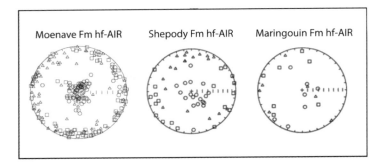

Fig. 2.3 High field anisotropy of isothermal remanence (hf-AIR) for red bed formations. From left to right: the Triassic Moenave Formation from the Colorado Plateau and the Carboniferous Shepody and Maringouin Formations from Nova Scotia. The minimum principal axes (circles) are near to perpendicular to bedding and the maximum (squares) and intermediate axes (triangles) are near to the bedding plane in these equal-area nets. The remanence anisotropy shows the fabric of the ChRM-carrying hematite particles. Figures from Bilardello & Kodama (2009b) and McCall & Kodama (2010).

compared to the foreset beds of cross-bedding in red beds, suggesting a DRM. This interpretation is based on early laboratory re-deposition experiments that showed the effect of initially dipping surfaces on the direction of the DRM acquired (King 1955; Griffiths *et al.* 1960). Others have used scanning electron microscopy (SEM) to examine the magnetic minerals in a red bed showing multiple generations of authigenic hematite (Walker *et al.* 1981). This result strongly supports a secondary chemical remanence for red beds. Recently, anisotropy of magnetic remanence fabrics we have collected from red beds show a strong depositional/compactional fabric with minimum axes clustered perpendicular to bedding and maximum and intermediate axes scattered in the horizontal (bedding plane) (Figs 2.3 and 5.9). This fabric is very similar to that observed for magnetite-bearing sedimentary rocks (Fig. 2.4; see also fabric for the Perforado Formation, Fig. 5.6 and the Pigeon Point Formation, Fig. 5.3) and strongly supports the notion that some, maybe most, red beds carry a DRM rather than a CRM.

Can traditional DRM theory explain the remanence of hematite-bearing rocks? The spontaneous magnetization of hematite is nearly 200 times less than that of magnetite, so is it possible that the torque of the geomagnetic field on hematite nano-particles would be reduced enough to preclude a prediction of perfect alignment with the field? Using reasonable values for the spontaneous magnetization of hematite, the viscosity of water and the geomagnetic field results in alignment times of 0.05 sec. This is still embarrassingly

short, and comparable to the alignment times for magnetite particles. In this calculation the magnetic moment of the nominally 1.0 μm diameter hematite particle is larger than that of the 1.0 μm magnetite grain in the earlier calculation because the hematite grain is assumed to be single domain rather than pseudo-single domain. This assumption is made because of hematite's low spontaneous magnetization and the importance of micro-crystalline anisotropy controlling remanence of a hematite particle (Butler 1992; Tauxe 2010).

$$t_0 = \frac{\pi d^3 \eta}{mB} = \frac{3.14159 \times 10^{-6} \times 10^{-3}}{1.2 \times 10^{-15} \times 50} = 0.05 \text{ sec}$$

Still, red beds typically have magnetizations about an order of magnitude stronger than magnetite-bearing sediments (10 mA/m for red beds versus about 1 mA/m for magnetite-bearing sediments), despite the fact that hematite has a much lower spontaneous magnetization than magnetite. This could suggest that hematite grains in a sedimentary rock are better aligned than magnetite grains.

Another way of investigating this question is to apply the approach used by Tauxe *et al.* (2006) in their re-deposition experiments. Tauxe *et al.* found a ratio of DRM: SIRM for re-deposited magnetite-bearing sediments that was of the order 10–30%, where SIRM is the saturation isothermal remanent magnetization that is applied to the rock sample in the laboratory. It is a measure of the total number of remanently mag-

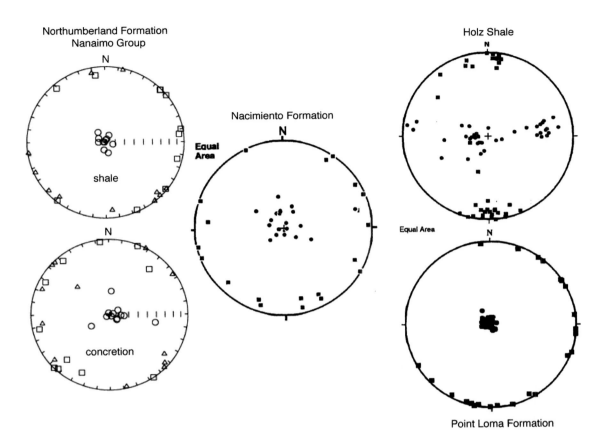

Fig. 2.4 Magnetic remanence fabrics for magnetite-bearing sedimentary rocks. The Northumberland Formation is part of the Cretaceous Nanaimo Group of British Columbia, the Paleocene Nacimiento Formation is on the Colorado Plateau in New Mexico and the Cretaceous Holz Shale of the Ladd and Williams Formation and the Point Loma Formation are from southern California. In all stereonet plots the minimum principal anisotropy of anhysteretic remanence (AAR) axes are perpendicular to the bedding plane and shown by circles (filled/unfilled); the maximum (squares) and intermediate (triangles) principal axes lie in the bedding plane. Figures from Kim & Kodama (2004) for the Northumberland Formation, from Kodama (1997) for the Nacimiento Formation and from Tan & Kodama (1998) for the Holz Shale and Point Loma Formations.

netized particles in the sample. In these experiments, Tauxe et al. provide evidence for flocculation controlling the DRM so this is the mechanism for reduction in DRM alignment from perfectly aligned grains (DRM saturation).

Tan et al. (2002) conducted re-deposition experiments with sediments reconstituted from the Eocene Suweiyi Formation red beds of central China. The purpose of Tan et al. was to see if synthetically induced depositional- and compaction-caused inclination shallowing in hematite-bearing rocks could be detected and corrected by a single component IRM applied at 45° to the vertical. However, Tan et al. also provide DRM: SIRM ratios for hematite-bearing sediments. Tan et al.'s hematite laboratory DRMs range 30–70 mA/m, while an IRM applied to this rock in a 1.3 T field is 6.5 A/m. Even though the rock is not magnetically saturated at this field, the IRM acquisition curves indicate it is close to saturation. The DRM: IRM ratios of 0.5–1.0% are much smaller than what Tauxe et al. observed; however, Tauxe et al. applied SIRMs to their sediments before re-deposition while Tan et al. did not.

For the proper comparison, Tan & Kodama (1998) also conducted re-deposition experiments with magnetite-bearing sedimentary rocks of the Cretaceous Holz Shale and the Point Loma Formation, again without applying an IRM to the sediments before re-deposition. These re-deposited sediments had DRM: SIRM ratios of 2.3% and 1.1%, respectively. The difference in alignment between re-deposited hematite-bearing sediments and magnetite-bearing sediments cannot be explained by the 200-fold difference in spontaneous magnetization of the different magnetic minerals, so a misaligning mechanism must also be affecting the DRM acquisition of hematite. Brownian motion is unlikely to affect detrital specular hematite particles that are typically several microns in size, so flocculation is the preferred mechanism. Since organic material is unlikely in the highly oxidized red sedimentary rocks, it is more likely that clay domains or van der Waals attraction of hematite to clay particles, as in magnetite-bearing rocks, reduces the alignment of the hematite grains.

THE ACCURACY OF A SYN-DEPOSITIONAL REMANENCE

The idea of a syn-depositional inclination error came from the first round of re-deposition experiments conducted by the first generation of paleomagnetists. In these experiments, typically varved glacial lake sediments were re-deposited in the laboratory in fresh water (King 1955; Griffiths et al. 1960). The original definition of the tangent–tangent relationship used to describe inclination shallowing resulted from this work:

$$\tan I_m = f \tan I_f$$

where I_m is the measured inclination of the re-deposited sediment, I_f is the laboratory magnetic field inclination and f is the flattening factor which lies within the range $0 < f \leq 1$. For the early re-deposition experiments, flattening factors of 0.4 were observed. The work of King and Griffiths et al. was also important because it showed that bedding slope and deposition from a current could affect the accuracy of the remanent magnetization acquired by the re-deposited sediments.

Subsequently, second-generation re-deposition experiments by Barton & McElhinny (1979) and Levi & Banerjee (1990) showed that slow re-deposition in the laboratory or directly in the cryogenic magnetometer showed no, or at least a very small (<3°), syn-depositional inclination error. It was then thought that the syn-depositional inclination error originally observed by King and Griffiths et al. was an artifact of their experimental procedure (Verosub 1977; Butler 1992). Measurements of natural sediments, particularly Opdyke & Henry's (1969) study of recent marine sediments from the world oceans, showed no inclination error. We have also shown (Chapter 1) that many recent lake and marine sediments appear to give accurate records of the geomagnetic field.

How can we explain the King results, the slower second-generation re-deposition experiments and the observations of an apparently accurate DRM in recent natural sediments? The prevailing idea at the time (Verosub 1977) was that, even if the magnetic particles initially had an inclination error at touchdown, they would quickly reorient themselves with the Earth's magnetic field in an early post-depositional remanent magnetization (pDRM) that accurately recorded the direction of the geomagnetic field. pDRM will be discussed in more detail in the next chapter. It is interesting to note that, in the recent re-deposition experiments and flocculation models of DRM acquisition, post-depositional remanence is not considered important because the magnetic particles are trapped in flocs and cannot easily reorient after deposition. Indirect evidence from Sun and Kodama's laboratory compaction experiments also indicates that the magnetic particles are locked onto clay particles at deposition by electrostatic or van der Waals forces, and are essentially immobilized.

Can the results of the first- and second-generation re-deposition experiments be explained by the flocculation model of DRM acquisition? In all the re-deposition studies, the salinity of the water is not explicitly mentioned but they all use freshwater sediments: King (1955) and Griffiths et al. (1960) used re-deposited glacial varves; Barton & McElhinny (1979) and Barton et al. (1980) used fine-grained organic-rich lake sediments; and Levi & Banerjee (1990) used lake sediments from Lake St Croix in Minnesota.

In the Griffiths et al. (1960) study, re-deposition experiments were conducted with different sediment grain sizes of c. 2–30 μm. All the experiments showed roughly the same magnitude inclination error (10–20°). The grain size and the sediment type (glacial varves) would suggest that the sediment was mainly quartz grains, so flocculation is unlikely for these experiments. The experiments by Barton & McElhinny (1979) and Levi & Banerjee (1990) were conducted with lake sediment that was described as organic-rich mud or as silty mud that had about 10% organics and 50% clay content (Lake St Croix sediments). It is highly likely that organic floccules were present in these re-deposition experiments while clay floccules are less likely to have been present because fresh water was most likely used.

Only the Mitra & Tauxe (2009) experiments studied inclination error and varied the salinity of the water. The synthetic sediments were composed of clay and maghemite, and settled over 2 weeks. The lowest salinity results reported (about 2 ppt) showed inclination shallowing with an f factor of about 0.5 (field inclination c. 45°, measured inclination c. 27°). Levi and Banerjee's slower experiments, which showed small inclination errors ($f = 0.93$), were conducted over 'several' weeks. Barton and McElhinny's experiment, conducted over 9 months, showed no inclination error. The initial explanation for the shallow inclinations in the King and Griffiths *et al.* experiments, that the re-deposition occurred too quickly for the particles to become aligned, is probably correct. From the experimental data in the literature, for freshwater laboratory deposition the longer the deposition time the smaller the inclination error. This interpretation can also explain results for sediments that are likely to contain organic floccules.

LABORATORY RE-DEPOSITION EXPERIMENTS FOR HEMATITE-BEARING SEDIMENTS

There have only been a few laboratory re-deposition studies using hematite-bearing sedimentary rocks, i.e. red beds, probably because the overwhelming bias among paleomagnetists is that red beds carry a secondary CRM rather than a DRM. However, the evidence from the anisotropy results for red beds indicate that the possibility of a red bed DRM should be taken seriously. Tauxe & Kent (1984) conducted a pioneering set of re-deposition experiments. They investigated the natural re-deposition of Siwalik Formation red bed sediment after flood events along the Soan River in Pakistan and also laboratory re-deposition of the same material. Tauxe and Kent observed significant inclination shallowing in the laboratory ($f = 0.55$) and in the natural re-deposition of Siwalik material along the

Soan River ($f = 0.4$). The declination of the re-deposited sediment remained accurate. The laboratory re-deposition was conducted in fresh water to minimize flocculation and, of course, the Soan River re-deposition also occurred in fresh water. Furthermore, both re-deposition experiments occurred in a short time, the Soan River sediments were deposited 'during the rains two weeks prior to sampling' and the laboratory re-deposited sediment was allowed to settle overnight (or until the settling tubes became clear) i.e. about 5 hours. These results support the interpretation that the syn-depositional inclination error is due to rapid deposition, and it may be eradicated by very slow deposition or by realignment of magnetic particles during the acquisition of a post-depositional remanence. Furthermore, Tauxe and Kent showed that their re-deposited sediments could acquire a pDRM and that it accurately recorded the field.

Tan *et al.* (2002) conducted laboratory re-deposition experiments with material obtained by breaking down Eocene Suweiyi Formation red beds from central Asia. They ultra-sonicated the red beds to liberate the magnetic mineral (hematite) from the non-magnetic matrix grains, added the material to distilled water to make a slurry and let the slurry settle in a laboratory magnetic field with an inclination of 58°. Unexpectedly, the slurry spontaneously segregated into two different grain size fractions: fine and coarse. The fine-grained slurry, whose paleomagnetism is apparently carried by pigmentary hematite grains, acquired an accurate record of the laboratory field direction after several hours of settling, and was then compacted with volume losses up to 70%. Their inclination was subsequently shallowed by compaction ($f = 0.52$). In the re-deposition experiments with the coarse-grained fraction, significant syn-depositional shallowing was observed that ranged from as much as $f = 0.34$ but as little as $f = 0.96$. For two of the eight coarse-grained sediment re-deposition experiments, inclinations actually steepened ($f = 1.10$ and $f = 1.38$).

One critical observation from this work is that the faster the deposition rate, the shallower the inclination at deposition. Significantly, coarse-grained sediments lost little inclination when compacted, probably because volume losses of only 10% could be obtained at the same pressures that caused 70% volume loss in the fine-grained sediments. Tan *et al.* (2002) conclude that the occurrence of an inclination error depends on the ability of hematite grains to reorient post-depositionally. Coarse-grained sediments are less likely to have this ability, particularly if the sediment accumulation rate is high. Fine-grained sediments can more easily acquire an accurate post-depositional remanence, but they are particularly susceptible to compaction-caused inclination shallowing while coarse sediments are not. It turns out that both coarse-grained and fine-grained hematite-bearing sedimentary rocks will likely have inclination shallowing, but they will acquire the shallowing at different times (either at or after deposition).

For completeness, two other early re-deposition experiments conducted with hematite-bearing sediments should be mentioned. Lovlie and Torsvik (1984) re-deposited naturally disaggregated Devonian Wood Bay Formation red bed sediment collected from recent floodplain deposits. They re-deposited the sediments in fresh water over a period of 5–7 days, then allowed the sediment to settle for an additional 4–6 days. What is significant about their study is that they conducted the re-deposition in a range of field inclinations and were able to observe the tangent–tangent relationship initially proposed by King (1955). The inclinations of the re-deposited red bed material had inclination shallowing with $f = 0.4$, but there was a good deal of scatter in their results. The other interesting observation from their study is that deposition and settling over 1–2 weeks did not eliminate inclination shallowing. This is consistent with experiments conducted with magnetite in fresh water discussed in the previous section. Lovlie & Torsvik (1984) also measured the anisotropy of magnetic susceptibility (AMS) of the re-deposited sediments and saw a typical oblate depositional fabric with vertical minimum principal axes and maximum principal axes scattered in the horizontal. The earliest hematite re-deposition experiment in the literature is Bressler & Elston's (1980) re-deposition of sedimentary material disaggregated from the Triassic Moenkopi red beds. They observed inclination shallowing but not the King (1955) tangent–tangent relationship between remanent inclination and field inclination.

DEPOSITION ON A SLOPING BED AND FROM A CURRENT

Early re-deposition experiments showed that initial bedding slope and deposition from moving water could affect the accuracy of a syn-depositional DRM (Fig. 2.5). Verosub (1977) provides a very useful review of

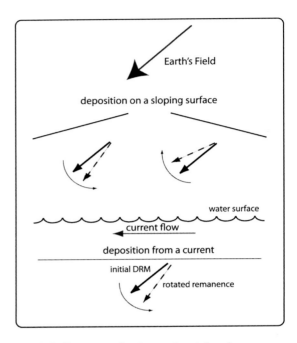

Fig. 2.5 Illustration of early experimental work investigating deposition on a sloping surface (15° in figure) or from a water current. The Earth's field inclination above is set at 40°. The black solid arrows show an accurate DRM in the sediment. The dashed arrows show the result of rotation caused either by deposition on a sloping surface (above) or from a current (below). The amount of rotation is 15° in the figure. Rotations are more complicated when the water flow or slope is not in the magnetic meridian.

the observations and theory that resulted from these early experiments.

The bedding error studied by these early workers (notably King, Griffiths and Rees) was for small dips less than 10° and would be most applicable to the foreset beds of water-laid cross-bedding. From the re-deposition experiments a 'bedding error correction' was developed in which the magnitude of the bedding error was equal, or nearly equal, to the dip of the bed. King (1955) and Griffiths *et al.* (1960) looked at re-depositions in which the dip of the bed was in the magnetic meridian and showed a change in inclination. If the bed dipped in the direction of the magnetic field, the magnetic vector would steepen by approximately the amount of dip, although some experiments showed a grain-size dependence with the finest-grained sediments having the vector rotate by up to twice the amount of dip. In this case, if the initial dip of the bed was unknown to the paleomagnetist, the rotation of the paleomagnetic data into stratigraphic coordinates would essentially 'undo' the bedding error for this special case of beds dipping exactly in the direction of the magnetic meridian. However, King's (1955) re-deposition experiment showed that, for beds dipping opposite to the field direction, the inclination would shallow by about the dip of the bed. Griffiths *et al.* (1960) also looked at dipping natural sediments (glacial varves) and saw that the rotation of the vector due to initial dip was always around a horizontal axis parallel to the strike of the bed. In this more general case, the declination as well as the inclination of the remanence would be affected. In all cases the rotation could be explained as though the magnetic grains were rolling down the slope; the magnitude of the effect was restricted to the small slopes investigated (≤10°).

Deposition from moving water causes about the same magnitude directional error in the syn-depositional DRM as deposition on a sloping surface. The sense of the rotation of the DRM vector is around a horizontal axis perpendicular to the current direction and the vector rotates into the current. The early current deposition results by Griffiths *et al.* (1960) and Rees (1961) are complicated by the inclination error that was ubiquitous for the deposition of the Swedish varved material used almost exclusively in these experiments. For instance, in the flume studies of Rees (1961), the magnitude of the 'current error' depends on an assumption of an inclination error that varied over 10–23° for an initial field inclination of 67° (0.41 ≤ f ≤ 0.65). Based on this assumption, the 'current error' ranged over 6–15°. All of the sloping bed and current work was done with glacial varves or re-deposition of glacially varved material in fresh water, so the effects of flocculation are unknown. Before any general conclusions can be drawn about the size of current and bedding slope errors, work should be done with sedimentary material other than glacial varved sediments.

SUMMING UP: SYN-DEPOSITIONAL DRM

It is clear that although paleomagnetists have made good progress in understanding the primary magnetization process for sediments and sedimentary rocks, the mechanism of syn-depositional DRM is still being worked out. At the most fundamental level, it involves rotation of magnetic nano-particles into alignment with the ambient magnetic field. The typical intensities of DRM in natural rocks however shows that the magnetic nano-particles are not perfectly aligned with the magnetic field, although simple theory predicts it. Some kind of misaligning force needs to be postulated. As recently suggested by a series of workers, flocculation is a strong candidate.

Syn-depositional DRM can be quite accurate based on evidence from recent natural sediments, assuming that burial compaction has not yet occurred. In contrast to this, early re-deposition experiments do show a large inclination error, typically with $f = 0.4$ where

$$\tan I_{\mathrm{m}} = f \tan I_{\mathrm{f}},$$

leading to shallowing as great as 20° in the worst case. Very slow re-deposition experiments taking several weeks up to several months would suggest that the syn-depositional inclination error could be mitigated by post-depositional reorientation of the magnetic nano-particles. Deposition on a sloping surface or from moving water has been shown to deflect the remanence from the ambient field by up to 10°; however, these results are based on very specialized conditions (Swedish varved material) and it is therefore difficult to know if the results can be applied more generally.

Post-Depositional Remanent Magnetization

One of the conundrums brought on by the early DRM re-deposition experiments was how to explain the large inclination error found in laboratory re-deposited sediments when paleomagnetists sensed that at least recent sediments were accurate recorders of the geomagnetic field. This hunch was supported by a classic 1969 paper by Opdyke & Henry (1969) that looked at the paleomagnetism of young marine sediments over a range of latitudes. The inclination of these sediments fit the predicted inclination for their latitude, not only supporting a geocentric axial dipole but accurate sedimentary paleomagnetism and no, or little, inclination shallowing.

Ted Irving, one of the important early contributors to paleomagnetism, had already proposed in 1957 (Irving 1957) a way in which sediments could accurately record the geomagnetic field despite the results of the re-deposition experiments, i.e. the small magnetic mineral grains in a rock were mobile when initially deposited and could reorient parallel to the geomagnetic field after deposition. He based his idea on the observation of a uniform magnetization in soft-sediment slumps in the Torridonian sandstone, indicating that the sediments had to have acquired their magnetization after slumping. He followed this proposal with an experimental result (Irving & Major 1964) in which a dry silt-sandstone/magnetite mixture, flooded with water in the presence of a range of magnetic field inclinations in the laboratory, accurately recorded the laboratory magnetic field.

Thus was born the concept of a post-depositional remanent magnetization (pDRM) in its simplest form. A pDRM not only explains accurate sedimentary records of the geomagnetic field, but has other implications for the interpretation of the paleomagnetism of sedimentary rocks. The main implication is that, if pDRM is the way sediments are typically magnetized, a sediment's paleomagnetism is younger than the depositional age of the sediment. To complicate matters, further down in the sediment column the sediment's magnetization will appear older than its depositional age. A second implication, that comes from the theoretical modeling of the acquisition of a pDRM, is that the geomagnetic field variations recorded by a sediment will be smoothed if different sub-populations of the magnetic mineral grains in a rock acquire their magnetization at different times after deposition.

Irving's original proposal of a pDRM was particularly far-sighted, given that very early paleomagnetic results from natural varved sediments from Scandinavia showed magnetizations pre-dating soft sediment deformation slumps and also had anomalously shallow inclinations (e.g. Johnson *et al.* 1948; Granar 1958).

Paleomagnetism of Sedimentary Rocks: Process and Interpretation, First Edition. Kenneth P. Kodama.
© 2012 Kenneth P. Kodama. Published 2012 by Blackwell Publishing Ltd.

Ken Verosub published an important review paper on sedimentary paleomagnetism in 1977. In the paper, Verosub made a huge conceptual leap about the paleomagnetism of sedimentary rocks. If deep-sea marine sediments carry an accurate record of the geomagnetic field, as exemplified by the study of Opdyke & Henry (1969), then they must have acquired their remanence by post-depositional remanent magnetization. This conclusion makes sense if you assume all syn-depositional remanence is inaccurate (as the very early re-deposition experiments seemed to show), but doesn't follow from the later very slow re-deposition experiments discussed in Chapter 2.

Verosub goes on to suggest that the mechanism for a pDRM, for marine sediments at least, is bioturbation. There was contradictory evidence in the literature when Verosub proposed this pDRM mechanism. Some workers (Keen 1963; Harrison 1966) stated that bioturbation would destroy a sediment's initial DRM. Other workers investigated the nature and accuracy of a pDRM caused by post-depositional disturbance with different illustrative experiments. Kent (1973) conducted a simple stirring of high-water-content marine sediments in the laboratory to show that bioturbation could reset the magnetization of a sediment and accurately record the direction of the geomagnetic field over a range of inclinations. Tucker (1980) followed up with 'stirred remanent magnetization' experiments and a more sophisticated laboratory set-up than that used by Kent (1973), but came to essentially the same conclusion: stirring, a model of bioturbation, could reset the magnetization of a sediment with a magnetization proportional in strength to the ambient magnetic field. Based on these results, a stirring remanence has been the starting point for compaction/inclination shallowing laboratory experiments (Deamer & Kodama 1990; Sun & Kodama 1992), assuming it is the correct model for an accurate marine sedimentary paleomagnetism.

Steve Graham (1974) did one of the first *in situ* bioturbation/DRM experiments by measuring the magnetization of tidal flat sediments from a heavily bioturbated region of San Francisco Bay. The sediment's magnetization accurately recorded the direction of the local geomagnetic field. The chemistry of the sediments, the evidence of heavy bioturbation and subsequent laboratory re-deposition experiments with the sediment convinced him that bioturbation caused the sediments to acquire an accurate pDRM.

Brooks Ellwood (1984) reported another field experiment investigating bioturbation as a mechanism for pDRM acquisition. He sprinkled a 1 mm thick layer of magnetite in a trench dug in tidal flat sediments on Sapelo Island, Georgia in a zone of active bioturbation. He then measured the sediment's magnetization periodically. It took 50 days for the magnetization to stabilize in the sediments, but the magnetization was 20° shallower than the local magnetic field. After 211 days and intense bioturbation during the summer, the sediment's magnetization began to approach the magnetization of the local geomagnetic field. Ellwood concluded that bioturbation does not reset a sediment's magnetization unless it is very intense. He notes that the bioturbation in the tidal flats of Sapelo Island is probably more intense than in deep-sea marine sediments.

In contrast to this is the laboratory experimental work conducted by Katari et al. (2000) in which natural marine sediments were allowed to be bioturbated by polycheate worms introduced to the sediments in the laboratory. Despite active and obvious bioturbation for three weeks, the magnetization of the sediments was not reset unless the sediment had been resuspended in a fecal mound. From this, Katari et al. concluded that bioturbation cannot easily reset a sedimentary magnetization and is therefore not an important cause of a pDRM. In fact, we see that in subsequent investigations into pDRM Tauxe and her colleagues (Katari et al. 2000) conclude that pDRM is not important in causing the magnetization of natural sediments.

The conflicting results from the different bioturbation and stirring experiments cited here may be due to the difference in time that bioturbation was allowed to act on the sediments, i.e. 3 weeks versus 211 days or potentially much longer (but undetermined) in Graham's (1974) experiment. However, Tauxe and her colleagues would suggest another explanation: many of the pDRM experiments were flawed because the natural sedimentary fabric had been destroyed, i.e. by stirring or experiment preparation (digging a trench) or drying before measurement, and that an undisturbed fabric comprising flocculated organics and clay grains typically traps and immobilizes the magnetic mineral grains directly at deposition. Payne & Verosub (1982) had come to the same conclusion earlier based on experiments in which samples, initially with high water content, were turned in a magnetic field at different points of drying. For samples with less than 60% sand, they found that the samples did not acquire any pDRM. Stober & Thompson's (1979) re-deposition experiments with Finnish lake sediments also led them to conclude that magnetic grains were immobilized soon after deposition by the formation of organic gels.

Perhaps the simplest explanation is that the absence/presence of a pDRM depends mostly on lithology. Paleomagnetists typically target the finest-grained sediments because that would ensure the quietest depositional environment and the least disturbed sediment. The finest-grained sediments would have the highest clay content and possibly a high organic fraction, thus a strong sedimentary fabric immobilizing the magnetic particles. As sediment grain size increases and the fraction of quartz particles in a sediment increases, the pore spaces are larger allowing post-depositional magnetic particle movement and there is less chance that organic or clay floccules will trap the magnetic grains.

MECHANISM OF pDRM ACQUISITION

Although the dominance of a pDRM in sediments and sedimentary rocks may be controversial, it is worthwhile discussing some of the concepts in the literature about the acquisition of a pDRM and the observations that these concepts are based on. If the simple model of a pDRM is envisioned, that the magnetic particle grains are mobile after deposition and can re-orient parallel to the Earth's magnetic field, then the porosity of the sediment allows the mobility. As the sediment is buried more deeply in the sediment column the porosity decreases and, at some point in time after deposition and at a certain depth in the sediment column, the magnetic grains are immobilized and the pDRM is physically 'locked in'. The idea of a 'lock-in depth' for a pDRM originally came from an important laboratory experiment conducted by Lovlie (1974) in which sediments were incrementally deposited in a column over a period of time in the laboratory, and the declination of the ambient magnetic field was rotated by 180° halfway through the experiment. Sediments deposited 10 cm below the point at which the field was rotated recorded the new field direction.

Lovlie's experiments were followed by those of Hamano (1980) and Otofuji & Sasajima (1981). Both experiments showed basically the same behavior. Hamano compacted both synthetic and natural sediments slowly in the laboratory and turned on a magnetic field during a range of void ratios. Otofuji and Sasajima centrifuged sediment slurries for a range of sediment densities. In these experiments, the ability to acquire a post-depositional remanence decreases as void ratio decreases or sediment density increases. Hamano suggested that his results indicated that sediments acquired a pDRM at depths of 15 cm–2.5 m in the sediment column. Otofuji and Sasajima only reported sediment densities for the acquisition of a pDRM of 1.13–1.3 g/cc.

If this lock-in depth behavior was typical of natural sediments, then geomagnetic field behavior would occur lower in the sediment column or 'earlier' than the depositional age of the actual magnetic field changes. Furthermore, if the physical immobilization or lock-in of the grains was a function of the magnetic particle size and the size of the sediment's pore spaces, then larger magnetic particles would be locked-in at shallower depths in the sediment column than smaller magnetic particles. For a grain size distribution of magnetic particles, subpopulations of the magnetic particles would record the geomagnetic field at different depths in the sediment column, thereby smearing the record of geomagnetic field variations.

Denham & Chave (1982) presented a theoretical model of pDRM acquisition and lock-in depth in which the sediment is considered to be a fluid with viscosity which increases with depth; the characteristic alignment time of the magnetic particles in the sediment therefore increases exponentially with depth. This is the approach used in the modeling discussed in observations of natural sediments discussed later in this chapter. Katari et al. (2000) criticize this approach, based on evidence that natural sediments do not have a gradually increasing Newtonian viscosity with depth. Instead, they favor the approach of Shcherbakov & Shcherbakova (1983) who considered that sediments are better modeled with slow elastic strain and plastic deformation. Their slow elastic strain could allow post-depositional realignment, but no realignment could occur due to plastic strain. This theoretical approach better fits the results of laboratory experiments by Tucker (1980) and Katari et al. (2000).

Recent re-deposition experiments yielded conflicting views about lock-in depth and its thickness. Katari et al. (2000) conducted experiments with natural, undisturbed marine sediments and saw no evidence for realignment with a post-depositional field. Their work with three weeks of bioturbation in the laboratory also gave weak evidence for realignment of a sedimentary magnetization after deposition. Based on these results they present a model of deep-sea DRM acquisition in which a layer of mixing – a homogeneous layer of thickness L at the sediment–water interface – is where detrital magnetic particles are incorporated into the sediment column.

Directly below this is a thin layer dL; this is vanishingly thin for Katari *et al.* (2000), where pDRM realignment can occur. Below this are the historical layers that provide a stable record of geomagnetic field variations. In their modeling, Katari *et al.* assume a homogeneous layer about 2 cm thick. Carter-Stiglitz *et al.* (2006) have conducted laboratory re-deposition experiments that provide a similar, but different, picture of deep-sea DRM acquisition (Fig. 3.1). In their experiments, Carter-Stiglitz *et al.* 'froze' the magnetic particles in their re-deposited sediments in place by adding a gelatin to the sediment that solidified below 20°C. They could examine the degree of magnetization at different concentrations (or water contents) of sedi-

ment. Based on their work, they saw pDRM acquisition at concentrations $m_s/(m_s+M_{water})$ between 44 and 50% and very weak DRM efficiency at higher concentrations. They revisit Katari *et al.*'s model and indicate that the layer of pDRM acquisition dL could have a finite thickness. In their model, the lock-in depth would be dependent on the rate at which the sediment column is dewatered by burial compaction.

EVIDENCE OF pDRM IN NATURAL SEDIMENTS

Although laboratory re-deposition experiments can be very helpful, particularly because different conditions can be isolated and controlled, the best evidence for or against a significant post-depositional lock-in of sedimentary magnetization must come from observations of real sediments. The evidence of post-depositional lock-in can be divided into two general categories. First, there are observations of discrete offsets between the expected depth of some distinctive geomagnetic event and the observed depth of the paleomagnetic recording of the event. Almost all of the studies in the literature have used the observation of the Matuyama–Brunhes polarity transition boundary (MBB) in marine sediments, the most recent reversal of the geomagnetic field at 780 ka. The second type of evidence used to support a pDRM lock-in depth comes from modeling of the assumed smoothing of geomagnetic field variations by the lock-in of magnetization over a given depth (and magnetic grain size) range. We will take a quick look at the studies supporting (or not supporting) a lock-in depth in each category, and then criticisms of those studies.

de Menocal *et al.* (1990) is the first study to observe an offset in the MBB compared to its expected position with respect to a well-dated tie point in the oxygen isotope paleoclimate record in deep-sea marine sediment cores. The basic approach used was to compare the depth of the MBB in eight cores with the depth of the oxygen isotope interglacial stage 19.1 (Fig. 3.2). The robustness of the approach depends on the 8 cores having different sediment accumulation rates (which they do, varying over 1–8 cm/kyr). Essentially, the depth of the MBB with respect to oxygen isotope stage 19.1 is plotted as a function of sediment accumulation rate. The data fall on a straight line, but have a y-intercept from which the lock-in depth is determined. The slope of the line is the age difference between stage

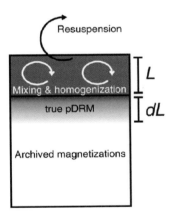

Fig. 3.1 Mixed layer L envisioned as due to bioturbation and zone of pDRM lock-in dL. The magnetization of the sediment is locked in at a depth $L + dL$ below the top of the sediment column and becomes part of the historical or archived record of the geomagnetic field. Figure is from Carter-Stiglitz *et al.* (2006) and represents the model of magnetization of Katari *et al.* (2000) for marine sediments. For Katari *et al.*, the lock-in layer dL is vanishingly small. For Carter-Stiglizt *et al.* and others the lock-in depth dL has a finite thickness. Some observations of natural sediments suggest it is 15 cm thick. Reprinted from *Earth & Planetary Science Letters*, 245, B Carter-Stiglitz *et al.*, Constraints on the acquisition of remanent magnetization in fine-grained sediments imposed by re-deposition experiments, 427–437, copyright 2006, with permission from Elsevier.

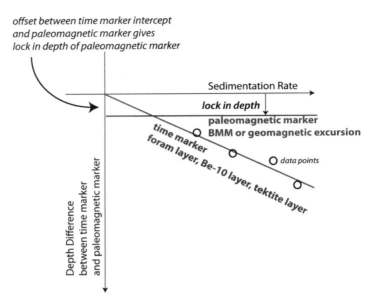

Fig. 3.2 Diagram of how de Menocal *et al.* (1990) and other workers determine the lock-in depth of a paleomagnetic marker such as the Brunhes–Matuayama polarity transition (BMM) or a geomagnetic excursion. Essentially, the depth difference between the paleomagnetic marker and an earlier time marker that doesn't have a lag in the sediment column (a foram layer, a ^{10}Be-dated layer or a tektite layer) is plotted versus the sediment accumulation rate. Each data point comes from a different core, sampling sediments with different accumulation rates. The *y*-intercept of the red line is a measure of the lock-in depth of the paleomagnetic marker. Figure redrafted from *Earth & Planetary Science Letters*, 99, PB de Menocal *et al.*, Depth of post-depositional remanence acquisition in deep-sea sediments: a case study of the Brunhes-Matuyama reversal and oxygen isotope Stage 19.1, 1–13, copyright 1990, with permission from Elsevier. (See Colour Plate 1)

19.1 and the MBB. There is another assumption thrown in, that bioturbation creates a mixing zone that affects the isotope record as well as the magnetic record, so the pDRM offset is in addition to the assumed 8 cm thick mixing zone caused by bioturbation. In the end, de Menocal *et al.* see a lock-in depth of 16 cm below the bioturbated mixing zone using this approach.

Tauxe *et al.* (1996) re-performed the de Menocal *et al.* study, but added 11 more cores for a total of 19 cores. Using their expanded dataset and without assuming a mixing model for bioturbation, Tauxe *et al.* saw that pDRM offset was 1 cm or less. The different results for the two studies could be partially due to the MBB being moved upcore in the Tauxe *et al.* study as the result of more rigorous alternating field demagnetization of the paleomagnetism in the sediments they added to de Menocal's dataset.

Channell *et al.* (2004) saw a shift of the position of the MBB to younger ages (by 4–5 kyr) in North Atlantic high-sediment-accumulation-rate cores with respect

to the calibrated age for the MBB. Channell *et al.* argue that the calibrated age of the MBB is based on its position in slow-sediment-accumulation-rate cores, and those cores are observed to have deeper lock-in depths than fast-accumulation-rate cores. This is an observation made in de Menocal *et al.*'s study, and would suggest that lock-in depth has affected the calibration of the MBB in slow-sediment-accumulation-rates cores and to a lesser extent in Channell *et al.*'s fast-sediment-accumulation-rate cores. Tauxe *et al.* (2006) criticize this interpretation, suggesting that the North Atlantic cores studied by Channell *et al.* are of drift deposits and their sediment has stayed in suspension in bottom currents for about 2 kyr before being deposited (not the 4–5 kyr offset Channell *et al.* observe), thus giving the too-young age for the sediments (since the age of the MBB in the core is based on the oxygen isotopic age of the sediment).

Knudsen *et al.* (2008) don't look at the position of the MBB, but at the depth of the paleomagnetic recording of a 190 ka geomagnetic excursion in Iceland Basin cores. They compare the depth of the directional changes of the excursion with the [10]Be record in the core. [10]Be production in the upper atmosphere is inversely proportional to the strength of the geomagnetic field, so the low geomagnetic intensities expected during a geomagnetic excursion should be matched with a peak in [10]Be production that is then deposited and recorded in the ocean's sediment. Knudsen *et al.* see no offset, suggesting no detectable lock-in depth.

Liu *et al.* (2008) essentially repeated the approach of de Menocal *et al.* (1990) and Tauxe *et al.* (1996), but limited their core-to-core comparisons to pairs of cores from the same oceanic basins. They considered water masses to differ from oceanic basin to basin, thus affecting the isotopic composition of the water and hence the isotopic composition of the shells made from the water. Since the depth of the fluctuations in the isotopic composition of the shells is ultimately what the depth of the MBB was being compared to, they felt that restricting the comparisons to a given basin would give more accurate results. When they used this approach they find an offset of the MBB by <20 cm from its true position with respect to the oxygen isotopic record; for Liu *et al.*, this offset however includes both the bioturbated mixing layer and the pDRM lock-in depth. They argue that most of the offset is due to mixing by bioturbation and not true lock-in depth, which they suggest may only be 1 cm or so thick.

Finally, Suganuma *et al.*'s (2010) study of the position of the MBB directional and paleointensity changes in marine sediment cores in the western Pacific compared to the [10]Be record of paleointensity variations seems to be clear-cut evidence of a 15 cm lock-in depth for the pDRM recorded by these deep-sea sediments. However, the match of paleomagnetic paleointensity variations and [10]Be variations is not straightforward and could affect the accuracy of the 15 cm offset determination. Furthermore, in Knudsen *et al.*'s (2008) study of the Iceland Basin, the sediment accumulation rate was nearly an order of magnitude higher (11–24 cm/kyr) than the western Pacific Ocean sediments studied by Suganuma *et al.* (2010) (<1 cm/kyr). These differences are one possible cause of the different results of these two similar studies. Suganuma *et al.* (2011) continued their pDRM lock-in studies with modeling of lock-in depth for both low-sediment-accumulation-rate cores from the Pacific (~1 cm/kyr) and high sedimentation rates from the North Atlantic

(~10 cm/kyr), and found evidence for a lock-in depth of 17 cm.

Lund & Keigwin (1994) is the first study to provide evidence of a pDRM lock-in depth by the 'smoothing of the paleomagnetic record' approach. The main evidence in support of a 10–20 cm lock-in depth comes from assuming that the marine sedimentary record of paleosecular variation of the geomagnetic field over the past 10,000 years has been smoothed with respect to lake sediment data recording paleosecular variation over the same period. The lake records used by Lund and Keigwin are from Minnesota and Great Britain. Their records are mathematically smoothed assuming 10 and 20 cm thick smoothing (lock-in) zones and then compared to the marine sediment paleomagnetic records from the Bermuda Rise measured by Lund and Keigwin. The comparisons are made for both inclination and declination.

Kent & Schneider (1995) compare three cores with records of the MBB and different sediment accumulation rates. The high-sediment-accumulation-rate cores show a double dip in paleointensity before and during the MBB, while the slow-sediment-accumulation-rate core shows only one dip. They show that they can reproduce the slow-sediment-accumulation-rate core dip record by sampling the record of the fast-accumulation-rate core at a slower rate, but also by assuming a pDRM lock-in depth of 16 cm.

Hartl & Tauxe (1996) obtain a different result with the same approach as Kent and Schneider, by adding five new studies to Kent and Schneider's study. Hartl and Tauxe show that detailed alternating field demagnetization and a more sophisticated determination of paleointensity in their records helps to resolve two dips in paleointensity, even in the low-sediment-accumulation-rate cores. They therefore argue that the need to assume a lock-in depth to explain only one dip in paleointensity in the slow-sediment-accumulation-rate cores is no longer needed.

These studies are presented in such detail in order to give a better sense of the quality and character of the evidence supporting, or not supporting, a finite lock in-depth; the major point is that all of the studies make assumptions and somewhat complicated analyses to observe lock-in depth. Tauxe *et al.* (2006) make some good criticisms of some of these studies, essentially repeating the arguments of Tauxe *et al.* (1996) and Hartl & Tauxe (1996), but we will examine their criticism of Lund and Keigwin in a little more detail simply because this study is the first and one of the best pieces

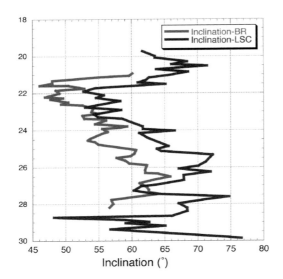
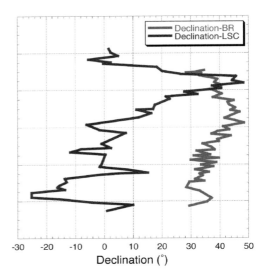

Fig. 3.3 Comparison of the Bermuda Rise (BR, Lund & Keigwin 1994) and Lake St Croix (LSC, Lund & Banerjee 1985) paleosecular variation records as envisioned by Tauxe *et al.* (2006). The inclination records (on the left) are correlated according to each core's age model as presented by Tauxe *et al.* (2006). The offset in absolute inclination is more than likely due to the difference in latitudes between the sites. As Tauxe *et al.* point out, the major features in the paleosecular variation show no offsets, arguing against a lock-in depth for the Bermuda Rise sediments. However, when declinations are compared for the two records (on the right) using exactly the same correlation, the declination swing to the east in the Bermuda Rise record is decreased in amplitude and increased in wavelength (smoothed) and offset down in the sediment column. This behavior is usually observed for a pDRM lock-in depth, in this case about 8 cm in the Bermuda Rise record. (See Colour Plate 2)

of evidence for smoothing of the paleomagnetic record of geomagnetic field variations by lock-in depth, thus having a serious effect on paleomagnetic accuracy (the theme of this book).

Tauxe *et al.* (2006) make the very good case that, in their comparison between the Bermuda Rise paleosecular variation record and the Lake St Croix PSV record, that Keigwin and Lund did not use the correct ages for the Bermuda Rise record. When the inclination records are compared using the correct ages there is no evidence of smoothing or any offset, thus arguing against any appreciable lock-in depth effects. A re-plot of their figure 9 supports the comparison made by Tauxe *et al.* (2006) of inclination records, but shows an absolute shift between the inclination records of about 5–8° with the Lake St Croix record being steeper.

This is most likely due to the 11.3° higher latitude of Lake St Croix compared to the Bermuda Rise site, thus inclination shallowing for the marine sediments does not need to be invoked to explain the difference. However, for a complete comparison, the declination

records should also be examined (Fig. 3.3). Both of the records show a shift of declination towards the east. The magnitude of the shift is much greater for Lake St Croix (70°) than for the Bermuda Rise (10°), but the shift occurs about 8 cm deeper in the marine record than it does for Lake St Croix. The wavelength of the eastern declination swing is broader and the amplitude attenuated in the marine record. The complete comparison of PSV records at the two sites therefore suggests no lock-in depth effects for inclination, but an 8 cm lock-in depth for declination with associated smoothing. It could be argued that the offset in declination is simply due to the westward drift of non-dipole magnetic field features (a well-known feature of the geomagnetic field's secular variation) and the Bermuda Rise's more easterly location with respect to Lake St Croix. The inclination records show no evidence of westward drift, however, so this explanation is probably not correct.

After all the pDRM studies have been considered, we have good evidence, experimental and observational, both for and against a post-depositional remanence with a 15 cm or so lock-in depth in marine sediments. The best way to proceed when considering the accuracy of marine sedimentary rocks is probably to assume that a pDRM could be present and that the age of magnetization could be offset from the depositional age of the sediments by some 2–15 kyr, given sediment accumulation rates of 1–8 cm/kyr. For most tectonic uses of paleomagnetism, this amount of offset is probably inconsequential. For faster sediment accumulation rates, for instance 10–100 cm/kyr for near-shore marine sediments, the pDRM could be locked in relatively soon (< 1 kyr) after deposition and only be smeared over hundreds of years (an inconsequential amount). For slower sediment accumulation rates (< 1 cm/kyr for deep-sea marine sediments), the pDRM could be locked in tens of thousands of years after deposition and the geomagnetic field record could be smeared over thousands of years, compromising the record of paleosecular variation.

If the paleomagnetism of a sedimentary rock is being used for magnetostratigraphy, the user should be aware that transition boundaries may be shifted slightly in time. This caveat is probably more important for marine sediments however, and not as much for the terrestrial sedimentary rocks often used for magneto-stratigraphic studies. There is very little observational evidence regarding pDRMs in terrestrial sediments but, even if it exists with the same magnitude of offset (10–15 cm), the higher sediment accumulation rates of terrestrial sediments would make the actual time offset between depositional and magnetic age smaller than in marine sediments. The smoothing that accompanies a lock-in depth of magnetization in slowly deposited marine sediments may attenuate the records of some high-frequency variations of the geomagnetic field, but could actually help in tectonic and magnetostratigraphic studies by partially averaging secular variation for a better resolution of the geocentric axial dipole field.

4

Inclination Shallowing in Sedimentary Rocks: Evidence, Mechanism and Cause

EARLY LABORATORY COMPACTION EXPERIMENTS

The classic paper by Opdyke & Henry (1969) showed clearly that recent marine sediments could accurately record the inclination of the geomagnetic field. This seemed to settle the question raised by the early laboratory re-deposition experiments suggesting that sedimentary rock and sediment inclinations were inaccurate and, as mentioned in the earlier chapter on DRM, the shallow inclinations in the re-deposition experiments were attributed to an artifact of the experiments. However, the Opdyke and Henry experiment did not consider the effects of burial compaction because the short gravity cores they measured were only a meter or so long; this is not deep enough to see the effects of the approximately 50% volume loss associated with burial compaction at depths of hundreds of meters.

Not all paleomagnetists were convinced that the inclination of sediments and sedimentary rocks was always accurate, particularly after compaction had affected the sediments. Blow & Hamilton (1978) conducted one of the earliest experiments designed to determine the effects of compaction on the directional accuracy of sediment. Their work with natural marine sediments did show that inclination could be flattened by dewatering of sediment. The initial inclination of 70° for their compaction experiment was decreased to 49° and the amount of shallowing could be explained by the volume loss of the sediment. However, Blow and Hamilton used evaporation of the sediment's pore fluid as a model of burial compaction. Evaporative drying is clearly not the mechanism by which deep-sea sediments are naturally dewatered. In fact, work by Noel (1980) soon after Blow and Hamilton's study suggested that drying alone could cause inclination shallowing due to surface tension forces in the pore spaces of the sediment. What was needed was a better model of natural dewatering to see how sediment inclinations would be affected.

In 1987, Anson & Kodama (1987) detailed an experiment designed to observe the effects of burial compaction on the inclination of marine sediment. They carefully dewatered their synthetic sediment in a laboratory compaction device initially designed by Hamano (1980) to study the lock-in of sedimentary DRM during the porosity decrease that occurs at the very top of the sediment column. Since Hamano only examined horizontal sediment magnetizations, he

Paleomagnetism of Sedimentary Rocks: Process and Interpretation, First Edition. Kenneth P. Kodama.
© 2012 Kenneth P. Kodama. Published 2012 by Blackwell Publishing Ltd.

could not determine whether a sediment's inclination changed during volume loss. The Hamano device worked by dripping water into a Plexiglas tank that vertically loaded a sediment slurry sample. The highest pressure that the device could reach was c. 0.15 MPa; higher pressures (up to 2.53 MPa) were achieved with a soil test consolidometer. Anson and Kodama used distilled pore water and kaolinite clay to mimic a natural sediment. They added both acicular and equidimensional magnetite to the clay to see the effects of magnetic particle shape on any inclination shallowing that they observed. The clay/magnetite slurries were compacted slowly in the laboratory so that the pore fluid could escape fast enough to avoid over-pressuring the samples, thus ensuring that the volume loss would mimic natural dewatering as closely as possible.

To magnetize the samples, the clay-magnetite-distilled water slurry was stirred in magnetic fields with inclinations 20–80° and intensities comparable to the Earth's magnetic field (nominally 50 μT). Stirring was used to mimic the acquisition of a pDRM, assumed at that time to be the mechanism of marine sediment magnetization (Kent 1973). Anson and Kodama also stirred a sample in a nearly zero intensity field (several nT) and found that the sediment acquired a magnetization at least two orders of magnitude smaller.

The main results of these experiments were that the samples lost 32–64% of their volume and inclination shallowed by c. 10–12° for initial inclinations close to 45°. Anson and Kodama observed a tangent–tangent relationship between initial inclination and compaction-shallowed inclination following Blow & Hamilton (1978) who followed King (1955):

$$\tan I_c = (1 - a\Delta V)\tan I_0$$

where I_c is the compacted inclination, ΔV is the volume lost, I_0 is the initial inclination and a is an empirical factor related to acicular or equi-dimensional magnetite, of value c. 0.55 for their experiments. For a more useful comparison to the other inclination-shallowing work described in this chapter, Anson and Kodama's experiments found $f = 0.61$ for acicular magnetite and $f = 0.70$ for equi-dimensional magnetite where $\tan I_f = f \tan I_0$ and f is the flattening factor of King (1955), I_f is final inclination and I_0 is the initial inclination.

Aside from the main result that volume loss like that experienced by natural sediments could cause significant inclination flattening, one of the main contributions from these experiments was Anson and Kodama's suggestion of a mechanism for compaction-caused inclination shallowing. They proposed that magnetite particles were sticking to clay particles by electrostatic attraction and were constrained to move with the clay particles as they rotated into the horizontal due to the vertical loading of the sediment slurry. This model was at odds with the ideas of the time which argued that, as sediment volume and its pore spaces decreased during compaction, the largest magnetic particles would be affected earlier than the smallest magnetic particles and thus should have greater amounts of inclination shallowing. Anson and Kodama's electrostatic sticking model predicted exactly the opposite behavior: the smallest magnetic particles would be the easiest to stick to clay particles and thus should experience the most shallowing. Alternating field demagnetization could separate the effect of compaction shallowing for large (low coercivity) and small (high coercivity) magnetite particles. In fact, alternating field demagnetization of the samples in Anson and Kodama's experiments showed that in two-thirds of the runs the smallest (highest coercivity) magnetite particles were flattened the most during compaction.

Deamer & Kodama's (1990) laboratory compaction experiments followed those of Anson and Kodama's. Deamer and Kodama's work was designed to test the 'electrostatic-sticking model' for inclination shallowing. In these experiments, four different single-clay synthetic sediments were compacted in the Hamano water tank consolidometer, as well as two natural marine sediments from off the coast of Oregon. Acicular and equi-dimensional magnetite of grain size 0.5 μm was also used for the single-clay synthetic sediments, as in the Anson and Kodama experiments. The main difference in the experiments was the use of saline pore waters to mimic natural marine sediments. In fact, Deamer and Kodama used Instant Ocean™ as the pore fluid. It is used for saltwater aquariums and has the same salts in the same proportion as ocean water.

Deamer and Kodama also varied the pH in their experiments to manipulate the electrostatic sticking effect. The pH of natural marine waters causes clay particles to have negative electric charges on their surface and for magnetite particles to have positive electric charges. The 'zero point of charge' for magnetite occurs at a pH of 6.5. For pH values < 6.5 magnetite has positive surfaces charges; for pH values > 6.5 it has negative charges. Altering the pH of the sediment slurry can either cause electrostratic attraction between magnetite particles and clay particles (for pH < 6.5) or repulsion (for pH > 6.5). In fact, Deamer and Kodama found that pushing the pH up greater than 6.5 had little effect on the amount of inclination flattening in their experiments, so the importance of the electrostatic sticking mechanism lost some traction.

Deamer and Kodama did show inclination shallowing due to compaction for a whole range of initial inclinations (initial inclinations of 30, 45, 60 and 75°), distilled and saline pore fluids, acicular and equidimensional magnetite, natural marine sediments and four different single-clay synthetic sediments (kaolinite, illite, montmorillonite and chlorite). The volume loss that occurred in their sediments (typically c. 55%) at the low pressures used (0.15 MPa) showed the same magnitude of inclination flattening observed in the Anson and Kodama experiments. The greatest amount of shallowing occurred at intermediate initial inclinations of 45–60° and was of the order 12°. The tangent–tangent relationship was not tested by Deamer and Kodama, but the plot depicted in Fig. 4.1 shows that the data clearly follow the tangent–tangent relationship of King (1955).

In Fig. 4.1, Deamer and Kodama's laboratory compaction data clearly fits the tangent–tangent relationship of King (1955) with the slope of the straight lines being the best-fit flattening factor. The data show that there is little difference in the flattening for natural sediments with saline pore fluid, single-clay synthetic sediments with distilled pore fluid and single-clay synthetic sediments with saline pore fluid. The f factors only vary over the range 0.68–0.74.

One interesting observation made by Deamer and Kodama is that the intensity of the sediment's magnetization decreased with compaction shallowing by 25–50%, an important consideration for relative paleointensity studies using marine sediments (Fig. 4.2). If the amount of intensity decrease is always roughly the same magnitude then relative paleointensity records from marine sediments would not be greatly affected; however, Deamer and Kodama observed that the greatest amount of intensity decrease occurred for the steepest initial inclinations and that there was a strong relationship between intensity decrease and initial inclination. In fact, Cogne (1987) saw hints of this effect when he deformed plasticine embedded with hematite particles with uniaxial compression. When the axis of compression was parallel to

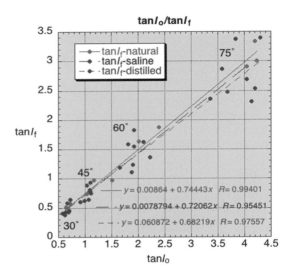

Fig. 4.1 tan I_f versus tan I_0 for single-clay synthetic sediments (kaolinite, illite, montmorillonite) in saline pore water (blue), distilled water (green) and for natural marine sediments (red). The slope of the lines indicates the best-fit flattening factor f for each sediment type. The f factors range from 0.68 for distilled pore water to 0.74 for natural sediments, very similar to those for Anson & Kodama's (1987) experiments. (See Colour Plate 3)

the magnetization it caused no change in the average magnetization direction for multiple subsamples, but caused an increase in scatter of the subsample magnetizations. If the same effect occurred at the magnetic particle scale, then compaction for steep initial inclinations should cause more misalignment of the magnetic grains and a greater intensity decrease than for samples with a shallower initial inclination.

A variable amount of intensity decrease with compaction could potentially have an effect on the relative paleointensity measurements made on marine sediments and would predict that intensity variations should be correlated to inclination variations for compacted sediments if the effect occurs in nature. Paleosecular variation (PSV) of the Earth's magnetic field causes variations in the local geomagnetic field inclination and declination on a time scale of thousands of years. These variations are due both to the variations in the non-dipole part of the geomagnetic field and to variations in the main geomagnetic dipole. If compaction causes an NRM intensity decrease that is a strong

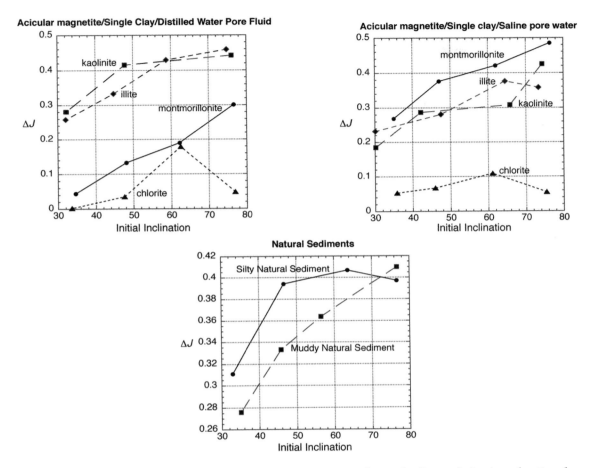

Fig. 4.2 Change in intensity (ΔJ) for single-clay synthetic sediments (top) and natural sediments (bottom) as a function of initial inclination during laboratory compaction. All synthetic sediments include 0.5 μm acicular magnetite and either distilled pore water (top left) or saline pore water (top right). Montmorillonite, illite, kaolinite and clay-rich natural sediments (Muddy Natural Sediment) show a strong relationship between initial inclination and the size of the intensity decrease. Decreased clay content in the natural marine sediment (Silty Marine Sediment) reduces the effect for steep and intermediate inclinations. (See Colour Plate 4)

function of initial inclination, then the resulting paleointensity record could have variations that are a function of PSV inclination variation and not entirely due to actual variations in the intensity of the geomagnetic field.

The possibility of a compaction effect on relative paleointensity records was checked by cross-correlating the paleointensity and inclination records for DSDP Site 522 from the South Atlantic Ocean (Tauxe & Hartl 1997). Only for depths between 70 m (7000 cm)

and 100 m (10000 cm) are there consistent anti-correlations between inclination and paleointensity, exactly what would be expected for a compaction effect on sediment magnetic intensity as observed by Deamer and Kodama in the laboratory (Fig. 4.3). At depths shallower than 70 m, the porosity of the sediments is >70% and hence not much volume loss due to compaction has occurred. At 70 m the porosity drops by 15%; compaction has therefore started to affect the sediments. The carbonate content of these sediments is

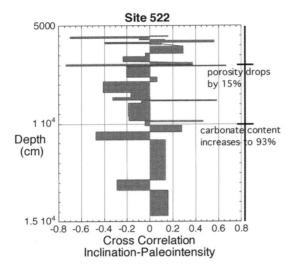

Fig. 4.3 Results of cross-correlation between Site 522 paleointensity results (Tauxe & Hartl 1997) and inclination record. Negative cross-correlations at depths between 7000 cm and 10 000 cm could be due to significant porosity loss at these depths and high enough clay content to make compaction effects on intensity apparent.

quite high and at 100 m depth increases to greater than 93%. We will show later in this section that clay content is an important factor controlling compaction-caused inclination shallowing. Clay content as low as 15% has been shown to cause inclination shallowing in Cretaceous marine sedimentary rocks from southern California (Tan & Kodama 1998). The clay content is *c.* 15% between depths of 60 and 100 m in the Site 522 core. The depth range of 70–100 m is perfect for observing if inclination is anti-correlated with paleointensity, and the results from this one core suggest that variable amounts of compaction-caused intensity decrease can affect the paleointensity record. In most cases however the effect is probably small or insignificant because of small variations in inclination and large genuine changes in paleointensity, which tend to minimize or swamp the compaction effect. The paleointensity record, stacked over the past 800 kyr (Guyodo & Valet 1999), shows good agreement over the world ocean suggesting that compaction has not appreciably affected these records.

Deamer and Kodama also pointed out that although their results did not support the electrostatic sticking

mechanism, it did show that significant shallowing occurred at low pressures and high porosities when the pore spaces would be larger than the 0.5 μm sized magnetite particles so some kind of clay-magnetite sticking mechanism was still probably occurring. They postulated that van der Waals forces could be responsible.

Sun & Kodama (1992) followed Deamer and Kodama's laboratory compaction experiments with more comprehensive experiments to test the mechanism for inclination shallowing, specifically the clay-magnetite sticking model. These experiments again used the Hamano water tank consolidometer to compact synthetic, single-clay sediments and natural marine sediments. The single-clay sediments used kaolinite and illite, 0.5 μm acicular magnetite and distilled water. The natural marine sediments contained about 50% clay content and had 2–3 μm natural magnetite. In the compaction experiments, saline pore water was used for the natural sediments. In all experiments at least 50% volume decrease occurred with up to 0.16 MPa of pressure. Inclination flattening and intensity decreases similar in magnitude to the previous experiments (Anson & Kodama 1987; Deamer & Kodama 1990) were observed, but some additional important observations were made. One of the most important came when the compacted and uncompacted sediments were critically point dried (to avoid surface tension effects during evaporative drying) and examined with a scanning electron microscope (SEM). These pictures clearly showed acicular magnetite grains sticking to clay particles in high-porosity sediments in which the pore spaces were much larger than the magnetite particles (Fig. 4.4).

In addition to this important observation were observations that provided a stronger hint about the microscopic mechanism of inclination shallowing. The inclination shallowing was not a simple linear function of void ratio (or volume) decrease. Most of the shallowing occurred at the lowest pressures (0.02–0.03 MPa), where the greatest volume losses occurred. During this period, volume loss occurred as pore spaces decreased in size with little reorientation of the clay grains making up the sediment. At higher pressures, inclination shallowing and volume loss occurred at a slower rate with increasing pressure. Here the mechanism of shallowing was the reorientation of clay grains into the horizontal with magnetite particles attached to clay grains and incorporated into clay domains. Clay domains develop in natural clay-rich sediments and are part of the clay fabric (Bennett *et al.* 1981). Clay

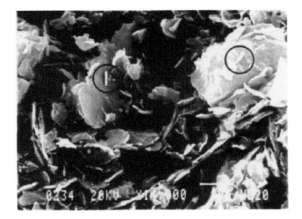

Fig. 4.4 SEM picture from Sun & Kodama (1992) showing acicular magnetite particles (circled) sticking to clay grains in high-porosity clay-rich sediment. W-W Sun and KP Kodama, Magnetic anisotropy, scanning electron microscopy, and x ray pole figure goniometry study of inclination shallowing in a compacting clay-rich sediment, *Journal of Geophysical Research*, 97, 19599–19615, 1992. Copyright 1992 American Geophysical Union. Reproduced by kind permission of American Geophysical Union.

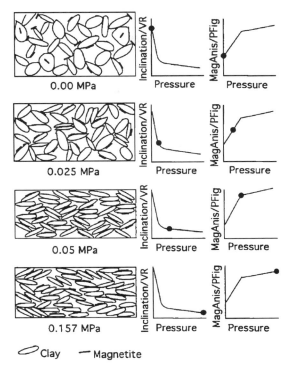

Fig. 4.5 Microscopic model for inclination shallowing from Sun & Kodama (1992) in which magnetite grains stick to clay grains in high-porosity sediments and are then incorporated into developing clay domains. The clay fabric is then rotated into the horizontal after the break in slope of the inclination shallowing versus pressure compaction curves. W-W Sun and KP Kodama, Magnetic anisotropy, scanning electron microscopy, and x ray pole figure goniometry study of inclination shallowing in a compacting clay-rich sediment, *Journal of Geophysical Research*, 97, 19599–19615, 1992. Copyright 1992 American Geophysical Union. Reproduced by kind permission of American Geophysical Union.

domains are similar to the flocs envisioned by Katari *et al.* (2000) to control the DRM acquisition of clay- and organic-rich sediments. Magnetic anisotropy and x-ray pole figure goniometry was also measured in Sun & Kodama's (1992) experiments, showing that both of these parameters also followed this two-part change during compaction. Sun and Kodama's model for inclination shallowing included a clay-magnetite sticking mechanism but also saw that incorporation into the clay fabric and the development of the clay fabric were important factors in shallowing, thus showing how the clay content of a sediment was critical for causing flattening (Fig. 4.5).

The natural sediments in Sun and Kodama's experiments had 50% clay content. Subsequent work by Tan & Kodama (1998) showed how the magnitude of inclination shallowing was directly related to the clay content of a sediment (Fig. 4.6).

One final point should be noted from the Sun & Kodama (1992) studies. Sun and Kodama argue that the porosities of their laboratory-compacted sediments suggest that inclination flattening occurs at burial depths of several hundred meters in the sediment column in marine sediments. The void ratios of their sediments indicate that their starting porosities are *c.*

80–85% for the kaolinite and natural marine sediments, consistent with initial porosities of sediments collected by the Ocean Drilling Project (ODP). By pressures of 0.02 MPa in Sun and Kodama's experiments, essentially when the rate of volume loss and inclination-shallowing changes and the flat-lying clay fabric is first established, the porosity of the sediment is 67–76%. When compared to ODP results from similar composition sediments, this indicates a depth of burial of *c.* 200 m.

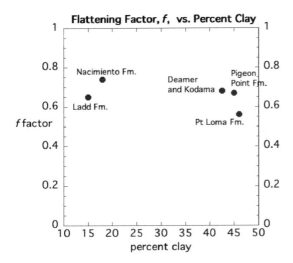

Fig. 4.6 Flattening factor $f = \tan I_f / \tan I_0$ for inclination shallowing as a function of percent clay content. The flattening factor decreases and inclination shallowing increases with increasing clay content. Clay contents as low as 15% are observed to cause significant inclination shallowing.

All of these observations made in the early laboratory compaction shallowing experiments at Lehigh could not have been made without the patient, careful and painstaking work of Gwen Anson, Gay Deamer and Wei Wei Sun.

LABORATORY EXPERIMENTS TO CORRECT INCLINATION SHALLOWING

Subsequent laboratory compaction work with magnetite-bearing sediments was all conducted as part of inclination-shallowing anisotropy-based corrections that will be covered in depth in the next chapter. This work was all published in the mid–late 1990s. Kodama & Davi (1995) corrected the inclination of the Cretaceous Pigeon Point Formation marine sediments from central California, Kodama (1997) tested the anisotropy-based inclination-shallowing correction with Paleocene continental sediments from New Mexico and Tan & Kodama (1998) conducted an inclination-shallowing correction of Cretaceous marine sediments from southern California. In all these experiments, the original sedimentary rock material was disaggregated and reconstituted as a slurry which was

then compacted in the laboratory using the Hamano device. The important advantage of using Hamano's device is that the load could be applied very slowly (over 8–10 hours) so that the slurry samples were not over-pressured causing non-equilibrium conditions, thus better mimicking natural burial compaction. In all these experiments the break in slope compaction behavior first observed in Deamer & Kodama's (1990) compaction experiments and then studied in detail by Sun & Kodama (1992) was also observed. In addition, inclination shallowing always occurred and the magnitude of the shallowing was similar (Pigeon Point Formation: $f = 0.67$; Ladd Formation: $f = 0.65$; Point Loma Formation: $f = 0.56$; Nacimiento Formation: $f = 0.74$) which, for the intermediate inclinations for these rocks, resulted in c. 10–20° of inclination shallowing.

Tan *et al.* (2002) was the first paper to examine the effects of laboratory compaction on the inclination of red bed analogues with natural hematite particles carrying the remanence of the sediments. Tauxe & Kent (1984) conducted a pioneering study of inclination shallowing in red bed material collected from the Siwaliks of Pakistan, re-deposited both in the laboratory and in natural conditions along the floodplains of the Soan River. Their re-deposition experiments showed large amounts of syn-depositional inclination shallowing with a flattening factor of $f = 0.4$. Tan *et al.* (2002) however re-deposited red bed material in the laboratory and then compacted it with the Hamano compaction device to see the effects of post-depositional compaction. The material they used was disaggregated from the Eocene Suweiyi Formation of central Asia, one of the units showing anomalously shallow inclinations for this part of the world during the Late Cretaceous–Early Tertiary.

To reconstitute these red beds, distilled water was added to make a slurry for the compaction experiments. One complication that occurred during the work was that the slurry spontaneously segregated into silt- and clay-sized parts. These two different slurries were compacted in the laboratory as well as a 1 : 1 mixture of the two. The clay-sized slurry experienced 17–19° of inclination shallowing ($f = 0.52$) while the silt-sized slurry showed little laboratory compaction inclination shallowing (3° or $f = 0.89$). The 1 : 1 mixture of clay and silt material also saw small amounts of shallowing (6° or $f = 0.80$). The break in slope compaction behavior first observed by Deamer and Kodama was also observed, particularly for the clay-sized slurry, reinforcing the notion that clay-rich sediment and the establishment of a clay fabric during compaction is the reason for this behavior.

The observation of this behavior is significant for understanding inclination shallowing because of its prevalence in all the laboratory compaction experiments. The simplest explanation for the disparity in the amount of inclination shallowing for the different grain-size slurries is that the amount of inclination shallowing is proportional to the amount of volume lost during compaction. The clay-sized fraction experienced up to 75% volume loss, the coarse-grained sediments only experienced about 12% volume loss and the mixture experienced a 35% volume loss. Interestingly, simply dripping slurry into the sample holder used for laboratory compaction initially magnetized the samples. While the slurries experienced a small amount (6–8°) of initial syn-depositional inclination shallowing, this disappeared after the samples sat in the laboratory field for only about an hour; this suggests post-depositional realignment of the hematite grains. A separate re-deposition experiment with the coarse-grained slurry did show significant syn-depositional inclination shallowing similar to that seen by Tauxe & Kent (1984) if the sediment accumulation rates were fast (c. 5 mm/min). The fastest deposition rates experienced the most shallowing (28°), but slower rates saw little or no syn-depositional shallowing (0.5 mm/min).

The laboratory compaction work set the stage for work that was designed to search for inclination shallowing in natural sedimentary rocks. Paleomagnetists' experience with recent lake and marine sediments in the 1960s, 1970s and 1980s suggested that inclination flattening in sedimentary rocks was probably not common or important. The inclination error seen in the early re-deposition experiments that used mainly varved glacial lake sediments was attributed to being an artifact of the experiment's design. Either the fast sediment accumulation rate or the drying used to allow measurement of the re-deposited sediment was the suspected cause of the inclination error. Paleomagnetists also argued, as we saw in Chapter 3, that a post-depositional remanence would remove the effects of any syn-depositional inclination error even if a syn-depositional error sometimes occurred in nature. The laboratory compaction experiments showed that the 50% or larger volume loss at depths of hundreds of meters in the sediment column could shallow the inclination of both marine sediments and quite possibly a DRM or an early-acquired chemical remanent magnetization (CRM) in hematite-bearing red beds. At this point, workers started to look for inclination shallowing in sedimentary rocks.

OBSERVATIONS OF INCLINATION SHALLOWING IN SEDIMENTARY ROCKS

Magnetite-bearing sediments and rocks

Most of the observations of inclination shallowing in sediments and in sedimentary rocks were published in the 1990s and 2000s. The studies discussed here and listed in Table 4.1 are separate from the studies discussed in the next chapter that consider identification and correction of inclination shallowing by either the anisotropy technique or the elongation–inclination (EI) directional distribution technique (Tauxe & Kent 2004). Combining the two groups of studies gives a large dataset showing that inclination shallowing has occurred in both magnetite-bearing and hematite-bearing sediments and sedimentary rocks.

The studies in Table 4.1 that observed inclination shallowing in magnetite-bearing rocks can be divided into two basic groups: freshly cored marine and lake sediments and ancient sedimentary rocks. In the late 1980s, two important studies first unequivocally showed inclination shallowing in marine sediments downcore as porosity decreased due to burial compaction. Celaya & Clement (1988) suggested that carbonate content was the culprit and observed that the marine sediments with greater than 80% carbonate content had significant inclination shallowing ($f = 0.48$). Sager & Singleton (1989) saw significant shallowing ($f = 0.62$) in four out of ten cores drilled around a salt dome in the northern Gulf of Mexico. Arason & Levi (1990) documented inclination shallowing downcore in DSDP Hole 578 in the Pacific Ocean. The inclination shallows from 53° at the top of the core, an accurate direction for the core's latitude, by 6–8° at a depth of 120 m. Only a quarter of the shallowing could be attributed to Pacific Plate motion over the past 6.5 million years. The inclination shallowing ($f = 0.75$) can be correlated to a 3–4% porosity decrease downcore, strongly supporting the contention that compaction-caused inclination shallowing occurs in nature. McNeill (1997) saw evidence of inclination shallowing from a core into the Great Bahama Bank sediments. Some cores had very little shallowing ($f = 1.39$ and 0.92) while others had more significant shallowing ($f = 0.78$ and 0.84).

Table 4.1 Magnetite-bearing sediments and sedimentary rocks

Locality	Lithology	*f* factor	Reference
North Atlantic deep-sea sediments	Carbonate-rich (>80%) marine beds	0.48	Celaya & Clement (1988)
Gulf of Mexico marine sediments	Clay-rich marine sediments	0.62	Sager & Singleton (1989)
Pacific marine sediments, DSDP 578	Biosiliceous clay, marine sediments	0.75	Arason & Levi (1990)
Hawaiian lake sediments	Laminated silt and clay, lake sediments	0.82	Hagstrum & Champion (1995), Peng & King (1992)
Great Bahama Reef	Shallow marine carbonate sediments	0.84	McNeill (1997)
Dead Sea	Lake sediments	0.45	Marco *et al.* (1998)
Pleistocene tephra in Japan	Water-laid tephra	0.8	Iwaki & Hayashida (2003)
London Clay, Sheppey, England	Eocene mudstones	0.63	Ali *et al.* (2003)
Sicak Cermik geothermal field, Turkey	Bedded travertine	0.68	Piper *et al.* (2007)
Yezo Group, Cretaceous Japan	Marine sandstones and shales	0.71	Tamaki & Itoh (2008)
Yezo Group Cretaceous Japan	Marine sandstones and shales	0.71	Tamaki *et al.* (2008)
Early Cretaceous limestones, southern Alps	Marine limestones	0.89	Channell *et al.* (2010)
Donbas fold belt, Ukraine	Carboniferous marine limestones (uC/Permian red sandstones-not corrected)	0.65	Meijers *et al.* (2010a)
Average for Magnetite-bearing sediments and rocks		0.69 ± 0.13 *N* = 13	Bilardello & Kodama (2010b) *f* = 0.65 +0.14/–0.11, *N* = 9

Freshwater lake sediments have also been observed to enjoy significant inclination shallowing. Hagstrum & Champion (1995) were able to make a direct comparison of the inclinations recorded by lava flows and lake sediments in Hawaii over the past 4400 years. The expected geomagnetic axial dipole (GAD) field has an inclination of 35° and the lava flows recorded an average inclination of 32°; in contrast, the lake sediments had low inclinations of c. 27° (*f* = 0.73). Divergence between the lake sediment inclination and the expected inclination with depth in the lake core suggests that compaction is the cause.

Marco *et al.* (1998) studied the paleomagnetism of a core from Dead Sea lake sediments. At the bottom of the 27 m long core, inclinations were 22° shallower than the GAD field (*f* = 0.45). In the top 7 m of the core, however, the inclinations were steeper than the GAD field. The shallowing trend with depth again supports compaction as the cause of inclination shallowing in these saline lake sediments.

Ancient sedimentary rocks with magnetite also show evidence of inclination shallowing. Ali *et al.* (2003) saw shallowing in the Eocene mudstones of the London Clay Formation in Sheppey, England, with a magnitude *f* equal to 0.57–0.69 depending on which Eocene paleomagnetic pole provides the expected GAD field for comparison. Ali *et al.* also compared their results to the coeval and nearby marine sediments of DSDP Hole 550 and also observed shallowing in the marine sediments. Tamaki & Itoh (2008) and Tamaki *et al.* (2008) used the isothermal remanent magnetization (IRM) correction technique of Hodych & Buchan (1994; see next chapter for details of this technique) to identify inclination shallowing in the Cretaceous Yezo Group marine shales and sandstones from Hokkaido, Japan (*f* = 0.71). Their work indicates that the Yezo Group is still anomalously shallow after inclination correction, supporting 3400 km of northward motion since the Campanian. Channell *et al.* (2010) calibrated the magnetostratigraphy of Early Cretaceous limestones from the southern Alps to nannofossils and found inclination shallowing via the EI technique, but only a modest amount ($f = 0.89$).

The last example of inclination shallowing in magnetite-bearing rocks from the literature is Meijers *et al.* (2010a) who studied Carboniferous marine limestones and Permian red sandstones from the Donbas Fold Belt in the Ukraine. They wanted to help resolve the Pangea B problem, which holds that the overlap between the northern continents and southern continents in the Late Paleozoic caused by the paleomagnetic data is due to unrecognized inclination shallowing forcing the sampling localities closer to the equator in the paleogeographical reconstructions than they actually were (Rochette & Vandamme 2001). Meijers *et al.* (2010a) identified and corrected inclination shallowing with the EI technique in Carboniferous magnetite-bearing rocks. Inclinations steepened slightly but significantly with correction (average of $f = 0.65$). The upper Carboniferous–Permian red beds which Meijers *et al.* studied were not corrected because the results agreed with similar age results from the Donbas Fold Belt (Iosifidi *et al.* 2010) that were argued to be unaffected by shallowing. Also, the Permian results of Meijers *et al.* did not cause a Pangea B overlap, so were not suspect.

Two studies that do not fall into our wet sediment and dry rock categories should also be mentioned. Iwaki & Hayashida (2003) studied Pleistocene water-laid tephra deposits in central Japan. Anisotropy of magnetic susceptibility (AMS) and anisotropy of anhysteretic remanence (AAR), which will be discussed in the next chapter as a means of identifying inclination shallowing, were used to identify modest inclination shallowing in the tephra ($f = 0.8$). Piper *et al.* (2007) studied the paleomagnetism of travertine deposits from the Sicak Cermik geothermal field of Turkey and found that bedded travertine recorded paleosecular variation and the disturbance due to earthquakes, and had anomalously shallow inclinations ($f = 0.68$).

Table 4.1 shows that when the flattening factors for these 13 studies of magnetite-bearing sediments and sedimentary rocks are averaged, a flattening factor of 0.69 ± 0.13 results. Bilardello & Kodama (2010b) published an average flattening factor of $0.65 + 0.14/-0.11$ for nine magnetite-bearing rocks for which anisotropy-corrected inclinations were reported (different studies from those reported here). The agreement between the two average f factors shows that magnetite-

bearing rocks and sediments can be assumed to have inclination shallowing of typically f c. 0.65–0.69.

Hematite-bearing rocks

The reports of inclination shallowing in red sedimentary rocks (red beds) occurred somewhat later than with magnetite-bearing rocks. This delay was in no small part because of the controversy over how red beds are magnetized. Paleomagnetists have had long discussions about whether red beds acquire their magnetization by a DRM process, or by secondary growth of hematite grains sometime after deposition when they acquire a chemical remanent magnetization (CRM). This 'red bed controversy' consumed much of some paleomagnetists' time, particularly in the 1970s. The argument went that if the red bed carried a CRM it would be immune from the effects of syn-depositional inclination shallowing. This is a reasonable argument but, with the acceptance that burial compaction could shallow the inclination at burial depths of hundreds of meters, an early CRM would not necessarily be immune from compaction-caused inclination shallowing. The issue is still not completely resolved but the evidence from red bed anisotropy measurements (see Chapter 5) suggest that most red beds, particularly those that are demonstrated to provide reliable paleomagnetic measurements, have either been magnetized by a DRM or a very early CRM (early enough to be affected by burial compaction).

The first study that reports shallow inclinations in red beds, not counting the early re-deposition experiments with red beds covered in Chapter 2 (Lovlie & Torsvik 1984; Tauxe & Kent 1984), is Garces *et al.* (1996) who studied Miocene hematite-bearing sedimentary rocks exposed in the Neogene Catalan basins of Spain. They observed a correlation between AMS intensity and inclination shallowing that argued to them that the paleomagnetic inclinations were affected by compaction or syn-depositional gravitational forces. The greatest shallowing was observed in finely laminated siltstones that were from distal alluvial fan facies interbedded with transitional marine facies. Up to 40° of shallowing was observed in these rocks ($f = 0.21$). Other lithologies investigated by Garces *et al.* (including massive mudstones and breccias) had lower amounts of shallowing, typically 20° ($f = 0.48$).

The Mio-Pliocene Siwalik Group red beds of the Himalayan foreland have long been a paleomagnetic target because of their importance in helping to time the northward motion of India and its collision with Asia. Tauxe & Kent's (1984) early investigation into red bed inclination shallowing was with the red sediment collected from the Siwalik Group in Pakistan and re-deposited in the laboratory. Ojha *et al.*'s (2000) magnetostratigraphic study of the Siwaliks in Nepal observed 29.4° of shallowing for an expected inclination of about 50° ($f = 0.31$). Most of the paleomagnetic data were derived from laminated siltstones.

Perhaps the largest group of red bed studies that show inclination shallowing come from Cretaceous and Early Tertiary rocks of central Asia. Table 4.2 lists six studies from western China that report observations of a low inclination anomaly. The initial observations of the central Asian inclination anomaly in the 1990s were interpreted to indicate geologically unreasonable post-Tertiary northward motion of central Asia. Various scenarios were used to explain the data: some of the explanations included long-term non-dipole field contributions to the geocentric axial dipole or large geologically unrecognized shear zones. As inclination shallowing corrections were applied to the red rocks carrying the inclination anomaly (e.g. Tan *et al.* 2003) and sedimentary rock directions could be compared to local volcanic directions (Gilder *et al.* 2003; Tan *et al.* 2010), it became clear that inclination shallowing in red beds was the most likely explanation for the low inclination anomaly. One reason that inclination-shallowing anomalies were so glaring for these rocks is that the expected inclination for central Asia in the Cretaceous and Early Tertiary was close to 45°, where the $\tan I_f = f \tan I_0$ relationship will give the maximum amount of shallowing.

The studies showing inclination shallowing in red beds from western China include Oligo-Miocene red beds from Subei (Gilder *et al.* 2001), Miocene red beds from the Maza Tagh mountains in the Tarim Basin (Dupont-Nivet *et al.* 2002), Jurassic–Cretaceous red beds from western Xinjiang province (Gilder *et al.* 2003), Tertiary red beds (20–60 Ma) from the Hoh Xil basin of northern Tibet (Liu *et al.* 2003), Cretaceous red beds from southeast China (Wang & Yang 2007) and the Lhasa block (Tan *et al.* 2010). Tan *et al.*'s (2003) inclination-shallowing correction for the Cretaceous red beds of the Kapusaliang Formation in the northern Tarim Basin will be discussed in more detail in Chapter 5. Two studies of Cretaceous and Tertiary

Table 4.2 Hematite-bearing sedimentary rocks

Locality	Lithology	f factor	Reference
Neogene basins in Catalan, Spain	Miocene laminated red fine-grained sediments	0.21–0.48	Garces *et al.* (1996)
Siwalik Group, Nepal	Miocene red beds: siltstones, sandstones, paleosols	0.31	Ojha *et al.* (2000)
Subei, China	Oligo-Miocene red beds	0.49	Gilder *et al.* (2001)
Maza Tagh range, Tarim Basin, China	Miocene red beds	0.32	Dupont-Nivet *et al.* (2002)
Western Xinjiang, China	Jurassic–Cretaceous red beds	0.64	Gilder *et al.* (2003)
Hoh Xil basin, northern Tibet	Tertiary red beds; 20–60 Ma	0.47	Liu *et al.* (2003)
Sierra de los Barrientos, Argentina	Ediacaran red claystones	0.62	Rapalini (2006)
Jishui and Ganzhou, SE China	Cretaceous red beds	0.65	Wang & Yang (2007)
Cis-Urals, Russia	Permian–Triassic red beds	0.7	Taylor *et al.* (2009)
Lhasa block, China	Cretaceous red beds	0.49	Tan *et al.* (2010)
Ukraine	Permo/Carboniferous red and gray sedimentary rocks	0.58	Iosifidi *et al.* (2010)
Average for hematite		0.51 ± 0.14 $N = 11$	Bilardello & Kodama (2010b) $f = 0.59 +0.24/-0.19, N = 15$

Note: Three additional studies of hematite-bearing rocks did not find evidence of shallowing (see text).

volcanic rocks (Hankard *et al.* 2007; Dupont-Nivet *et al.* 2010) implicated inclination shallowing for sedimentary results from central Asia. In all these studies, the authors recognized that inclination shallowing had affected their results. Some of the later studies applied the EI correction to their results (Wang & Yang 2007; Tan *et al.* 2010). The average flattening factor of $f = 0.51$ for these six studies suggests inclination shallowing of about 25° for these rocks and paleolatitude anomalies of about 15°.

Inclination shallowing in red beds has been observed in other places outside of China. Taylor *et al.* (2009) observed shallowing in Permo-Triassic red beds from the Urals in Russia. Based on far-sided paleomagnetic poles for their data, they suggest inclination shallowing has affected their results. Iosifidi *et al.* (2010) detected it in Permo-Carboniferous red and gray sedimentary rocks in the Ukraine. In their results the component isolated at the highest demagnetization temperatures was shallower by 12° than the component removed at intermediate temperatures. The assumption they made is that the high-temperature component was carried by detrital hematite and affected by shallowing, while the intermediate temperature component was carried by a CRM in pigmentary hematite and unaffected by shallowing. This would predict a very low f factor for these rocks, so the f factor in Table 4.2 is derived from comparison of Iosifidi *et al.*'s results to the directions predicted from the accepted apparent polar wander path for this part of the world (Torsvik *et al.* 2008). Finally, Rapalini (2006) used AMS and IRM analysis to detect shallowing in Neoprotereozic (Ediacaran) red claystones from the Sierra de los Barrientos in Argentina.

Bilardello & Kodama (2010b) also summarized the f factors for hematite-bearing rocks for which inclination-shallowing corrections had been conducted. From a separate set of studies ($N = 15$), they found an average f factor of 0.59 which agreed nicely with the average f factor of 0.51 calculated for the 11 studies discussed here and listed in Table 4.2. Hematite-bearing rocks appear to be more affected by shallowing than magnetite-bearing rocks with shallowing of 25–30° for intermediate initial inclinations. This could be due to the combination of demonstrated syn-depositional shallowing (e.g. Tauxe & Kent 1984; Tan *et al.* 2002) and compaction-induced shallowing for finer-grained facies.

For completeness, we should point out that there are several studies of sedimentary rocks that explicitly looked for inclination shallowing and did not observe any. Smethurst & McEnroe (2003) compared Silurian volcanic and red bed directions from Newfoundland and saw no difference, although earlier studies (Stamatakos *et al.* 1994) did see evidence of shallowing. Rapalini *et al.* (2006) applied an IRM at 45° (the Hodych & Buchan 1994 correction technique) to Permian red beds from South America and saw no evidence of shallowing. Sun *et al.* (2006) compared directions from Cretaceous red beds and basalts from the Qaidam Block in China and saw no evidence of inclination shallowing in the red beds. Finally, Schmidt *et al.* (2009) performed a detailed IRM analysis of the Nucaleena Formation cap carbonate rocks from Australia and found no evidence of shallowing. What these studies show is that while inclination shallowing is much more common in rocks than originally thought, it is not present in absolutely every sedimentary rock. In the next chapter we will examine in detail techniques to identify and correct for inclination shallowing in sedimentary rocks.

5

How to Detect and Correct a Compaction-shallowed Inclination

As the evidence from laboratory experiments in the late 1980s and early 1990s began to mount that compaction could shallow the inclination of sediments at burial depths of hundreds of meters, the next logical step for paleomagnetists was to develop techniques that could discern if shallowing had occurred in particular sedimentary rocks and, if it had, to accurately correct for it. The first papers attempting a correction used an approach based on the magnetic anisotropy of sedimentary rocks. They are, in order of appearance, Hodych & Bijaksana (1993), Collombat *et al.* (1993) and Kodama & Davi (1995). Tauxe & Kent's (2004) elongation–inclination (EI) technique came along about a decade later and uses a totally different and independent approach for identifying and correcting inclination shallowing in sedimentary rocks. We will consider the EI technique later in this chapter after we have discussed the anisotropy technique; we will then compare the two techniques.

THEORETICAL BASIS OF THE MAGNETIC ANISOTROPY–INCLINATION CORRECTION

Magnetic anisotropy is simply a measure of the preferred alignment of magnetic particles in a rock. The bulk magnetic anisotropy of a rock sample is determined from measurements, made in different orientations, of the ease with which a rock or sediment sample is magnetized by an applied magnetic field. The measurements made in different orientations are fit to a second-rank tensor that describes the three-dimensional relationship between the field applied to the sample and the resulting magnetization (see Menke & Abbott 1990 for an excellent physically intuitive explanation of second-rank tensors). The tensor represents the magnetic anisotropy of the sample. Ultimately, the bulk anisotropy depends on two factors: the preferred alignment of the magnetic particles in the rock and the magnetic anisotropy of the individual magnetic particles in a rock. Without delving too deeply into the physics of the small (micron and submicron) magnetic particles that carry the paleomagnetism of sediments and sedimentary rocks, it is important to realize that a rock's magnetic particles have 'easy' directions along which they are typically magnetized, with a magnetic north pole at one end and a south pole at the other. For some magnetic minerals such as magnetite, the physical shape of the particle dictates the alignment of the magnetic 'easy' axis: it is parallel to the long axis of the particle. For other magnetic minerals such as hematite, the 'easy' magnetization axis is controlled by crystallography and typically

Paleomagnetism of Sedimentary Rocks: Process and Interpretation, First Edition. Kenneth P. Kodama.
© 2012 Kenneth P. Kodama. Published 2012 by Blackwell Publishing Ltd.

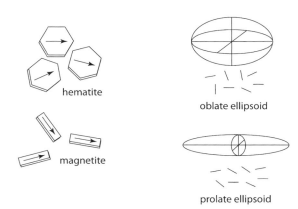

Fig. 5.1 Easy magnetic axes for hematite hexagonal plates (top left) and for elongate magnetite particles (bottom left). Oblate and prolate ellipsoids (right) graphically used to represent the orientation distributions of the easy axes of magnetic particles, shown by short lines.

lies in the basal crystallographic plane. The anisotropy of an individual particle is a measure of how easy it is to magnetize a particle along its 'easy' axis compared to being magnetized perpendicular to it (or along a particle's 'hard' axis).

Paleomagnetists characterize different kinds of magnetic anisotropy, also called magnetic fabrics, as either oblate or prolate because the second-rank tensor can be represented graphically by an ellipsoid. Oblate magnetic fabrics are represented by 'flattened' ellipsoids with magnetic particles lying closer to a plane (called the foliation plane), while prolate fabrics are 'lineated' with magnetic particles lying closer to a line (called the lineation). Sometimes pancakes are used to visualize oblate fabrics and cigars to represent prolate fabrics (Fig. 5.1). For sedimentary rocks affected by compaction, oblate fabrics lying parallel to the bedding plane are typically observed.

Jackson *et al.* (1991) is an influential paper that presented the theoretical basis for the use of magnetic anisotropy measurements to identify and correct shallowed inclinations in sedimentary rocks. Jackson *et al.* developed a theoretical relationship between the orientation distribution of magnetic particle easy axes and inclination shallowing. They assume that the orientation distribution can be measured by the magnetic anisotropy of a sedimentary rock. In the King (1955) relationship:

$$\tan I_c = f \tan I_0$$

where f is the flattening factor, I_c is the compacted inclination and I_0 is the initial, pre-compaction inclination, Jackson *et al.* have essentially shown that f is related to the bulk magnetic anisotropy and the individual particle anisotropy of the rock:

$$f = \frac{K_{min}(a+2)-1}{K_{max}(a+2)-1}$$

where K_{min} and K_{max} are the minimum and maximum axes of the ellipsoid used to represent the magnetic anisotropy (Fig. 5.1) and a is a factor that represents the anisotropy of the individual magnetic particles. The factor a is simply the ratio of the magnetization along the easy axis of an individual particle to the magnetization along the perpendicular, or hard, axis. One of the assumptions of Jackson *et al.*'s approach is that the anisotropy starts out very small or non-existent when sediment is first deposited and becomes increasingly flattened, i.e. oblate, with burial compaction. The inclination is also assumed to start out parallel to the ambient geomagnetic field at deposition and become shallower in a regular way as compaction proceeds. In Jackson *et al.*'s model there should be a direct relationship between the development of the preferred alignment of magnetic particle axes, the degree of magnetic anisotropy and the amount of inclination flattening.

Sun & Kodama's (1992) laboratory compaction experiments can be used to test these assumptions. One of the sediments compacted in Sun and Kodama's work was natural marine sediment containing natural magnetite with a grain size of 2–3 μm. Although the anisotropy of this sediment was not measured at 0 MPa of overburden, the pattern established between anisotropy and compaction at low pressures indicates that the anisotropy would extrapolate back near to 0 before compaction started (Fig. 5.2).

For this sediment the amount of shallowing at 0.15 MPa was 11.5° and the initial inclination 45°. Subsequent work done on natural marine sediments suggests that for natural magnetite the most typical a value observed is $a = 2$. Using these parameters in Jackson *et al.*'s correction equation gives an initial inclination of 45°, exactly what was measured in the laboratory at deposition. The most naturally realistic sediment compacted in the laboratory shows that the anisotropy–inclination shallowing correction

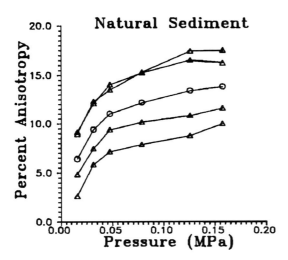

Fig. 5.2 Increase in magnetic anisotropy of natural marine sediment compacted in the laboratory by Sun & Kodama (1992). Although the magnetic anisotropy was not measured at 0 pressure, the extrapolation of the data back to 0 pressure indicates that the anisotropy was near to 0%. W-W Sun and KP Kodama, Magnetic anisotropy, scanning electron microscopy, and x ray pole figure goniometry study of inclination shallowing in a compacting clay-rich sediment, *Journal of Geophysics Research*, 97, B13, 19599–19615, 1992. Figure 7, page 19608. Copyright 1992 American Geophysical Union. Reproduced by kind permission of American Geophysical Union.

technique can give accurate results. The sediments in Sun and Kodama's experiments were all stirred to obtain an analogue of a pDRM acquired during bioturbation in marine sediments. Slow settling in a geomagnetic field may give an entirely different initial pre-compaction magnetic fabric. Recent work by Schwehr *et al.* (2006) supports low magnetic anisotropy at the deposition of natural sediment. They measured the anisotropy of magnetic susceptibility (AMS) of sediment downcore in the Santa Barbara basin and observed very low-intensity AMS fabrics for the surface sediments.

The assumptions in the Jackson *et al.* (1991) theoretical model for an anisotropy-based inclination correction could lead to overcorrection for an initial horizontally foliated fabric associated with an accurate syn-depositional inclination. However, the success of the anisotropy-based inclination correction (see

Kodama 1997, described below) and similar results from the EI correction technique of Tauxe & Kent (2004) and the anisotropy technique for the same rock units suggests that the commonly observed horizontally foliated fabric in natural sediments is probably always due to some amount of inclination shallowing, whether it is caused by compaction or mis-orientation of the magnetic grains at deposition.

EARLY INCLINATION-SHALLOWING CORRECTIONS

In the first inclination-shallowing study in the literature, Hodych & Bijaksana (1993) studied Cretaceous limestones collected from five equatorial Pacific DSDP cores. Hodych and Bijaksana closely followed the theoretical reasoning of Jackson *et al.* (1991) for an inclination-shallowing correction using magnetic anisotropy. The inclination of their samples was on average 17° shallower than the expected paleofield direction of 44° ($f = 0.53$). Magnetic anisotropy was measured using anhysteretic remanent magnetizations (ARM) applied to their samples in nine different orientations.

ARM is a laboratory-induced magnetization that can activate magnetic minerals such as magnetite with relatively low magnetic 'hardness' or coercivity. Magnetic hardness is a measure of how easy or how difficult it is to change the magnetization of a sample by the application of a magnetic field. Samples with high coercivities or high magnetic hardness need comparatively strong magnetic fields to alter their magnetization. ARM is a special manner of applying a magnetic field using a combination of alternating and direct magnetic fields. ARM is used in many experiments to measure the magnetic anisotropy of magnetite-bearing rocks.

Hodych & Bijaksana (1993) were able to correct the shallow magnetization of the limestone samples to the expected paleofield direction using the ARM magnetic anisotropy of their samples. Following Jackson *et al.*'s theoretical approach, they measured the individual particle anisotropy of their samples as well as the bulk anisotropy. Hodych and Bijaksana also looked at the degree of compaction of their samples based on the samples' porosity, and determined that compaction was the cause of the inclination flattening. Despite this pioneering effort of using Jackson *et al.*'s approach to correct inclination shallowing, Hodych did not pursue this line of research and only published one more inclination correction paper (Hodych *et al.* 1999) and a simplified version of the correction technique (Hodych & Buchan 1994) using an IRM applied at 45° to bedding. We will discuss the 45° IRM approach in more detail when we consider inclination-shallowing corrections for hematite-bearing rocks.

Collombat *et al.* (1993) also used magnetic anisotropy to study shallow inclinations in marine sediments, in this case recent marine sediments from the North American continental rise in the North Atlantic Ocean. They did not determine the complete magnetic anisotropy ellipsoid, as was done by Hodych & Bijaksana (1993), but averaged the ARM obtained in two perpendicular horizontal directions and divided that by the vertical ARM measured perpendicular to bedding, denoting it the H_a factor. The H_a factor was a relatively quickly measured approximation of the ARM second-rank tensor used in the Jackson *et al.* approach. In Collombat *et al.*'s study, H_a was found to be directly proportional to the amount of inclination shallowing downcore. Samples with little shallowing had very low anisotropy as measured by H_a, and samples with large amounts of shallowing (up to 30°) had H_a factors as large as 1.4. Collombat *et al.* (1993) also astutely pointed out that although susceptibility anisotropy (AMS) is easy to measure, they preferred to use ARM anisotropy for detecting inclination shallowing. This was because it is a direct measure of the magnetic particles carrying the paleomagnetism of the sediments, and would not be affected by clay particles and large magnetite particles that do not carry the ancient magnetism of the sediments. This realization foreshadowed the ARM anisotropy approach used by most inclination-shallowing studies over the next two decades.

Kodama & Davi (1995) conducted the first of a series of compaction-correction studies by the Lehigh paleomagnetics laboratory by studying the Cretaceous turbidites of the Pigeon Point Formation of northern California. The initial paleomagnetic results from the Pigeon Point Formation were reported by Champion *et al.* (1984) and suggested 21° of post-depositional latitudinal motion of the rocks. The Pigeon Point provided important evidence that California's Mesozoic tectonostratigraphic terranes had moved significantly northward since they formed. Although Champion *et al.* had considered the possibility that inclination shallowing had affected their results, they dismissed the idea. Their reasoning was essentially that slumped layers of rock in the Pigeon Point Formation had the same magnetic direction as flat-lying sedimentary layers, so the magnetization was a pDRM. They argued that pDRMs had been shown to be accurate and unaffected by inclination shallowing. However, they did not consider the possibility that compaction had flattened the Pigeon Point inclinations subsequent to deposition. Kodama and Davi therefore tested the possibility of compaction-caused inclination shallowing. They collected the mudstones and siltstones of the Pigeon Point, disaggregated them and compacted them in the laboratory, measuring both the decrease in inclination and the increase in ARM anisotropy during a volume loss up to 50%. They fit their inclination-anisotropy compaction results to Jackson *et al.*'s theoretical inclination-shallowing curves as a way of determining the appropriate individual particle anisotropy for an anisotropy–inclination correction. The ARM anisotropy of the natural Pigeon Point samples collected by Kodama and Davi indicated that the Pigeon Point had experienced about 10° of inclination shallowing ($f = 0.7$), not enough to explain all the shallowing (27°) used as evidence for paleolatitudinal offset. Even with an inclination-shallowing correction, the Pigeon Point Formation still indicated significant post-Cretaceous transport for the northern California tectonostratigraphic terranes.

One important observation made by Kodama & Davi (1995) was that the ARM anisotropy had a complicated magnetic fabric indicating a mixture of both oblate and prolate characteristics (Fig. 5.3). This made geological sense because the Pigeon Point rocks were deposited from turbidity flows. Turbidity flows had been shown to cause lineated fabrics, while the subsequent burial compaction after deposition would impose an oblate fabric. The correction, as envisioned by Kodama and Davi, would have slightly different results depending on whether an oblate or prolate fabric was used for the theoretical correction (12° for a prolate fabric; 8° for an oblate fabric). However, the composite ARM anisotropy for the Pigeon Point Formation does demonstrate that it is depositional and therefore primary, an important observation for any paleomagnetic study.

INCLINATION CORRECTIONS FOR TECTONOSTRATIGRAPHIC TERRANES OF WESTERN NORTH AMERICA

The Pigeon Point Formation study also showed inclination-shallowing errors of about 10° for the

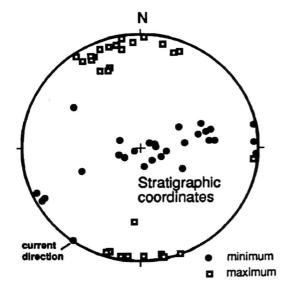

Fig. 5.3 A stereonet showing the magnetic fabric of the Cretaceous Pigeon Point Formation rocks from Kodama & Davi (1995). The minimum axes (solid dots) of the samples' magnetic fabric ellipsoids are scattered from vertical to horizontal while the maximum axes (open squares) are horizontal and oriented north–south. This orientation suggests a composite oblate and prolate fabric for the Pigeon Point resulting from both deposition from a turbidity flow current and from burial compaction. KP Kodama and JM Davi, A compaction correction for the paleomagnetism of the Cretaceous Pigeon Point Formation of California, *Tectonics*, 14, 5, 1153–1164, 1995. Figure 7, page 1160. Copyright 1995 American Geophysical Union. Reproduced by kind permission of American Geophysical Union.

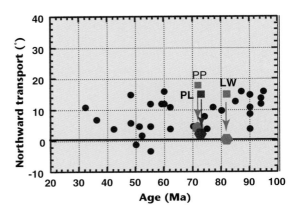

Fig. 5.4 Paleolatitudinal offset of California tectonostratigraphic terranes shown by red circles as a function of age in Ma. Offsets are calculated from the paleomagnetic data of sedimentary rocks for different tectonostratigraphic terranes compared to the predicted paleolatitude for cratonic North America for a particular age. Blue squares show the uncorrected paleolatitudinal offset for the Cretaceous Pigeon Point Formation (PP) and the Cretaceous Ladd and Williams Formation (LW). The blue arrows point to the corresponding anisotropy-corrected paleolatitudinal offsets shown by the blue hexagons. In these three cases, the paleomagnetic inclination for these formations was corrected for inclination shallowing with the remanence anisotropy measured for the rocks. (See references for the terranes depicted by the red circles in Butler *et al.* 1991.) (See Colour Plate 5)

Cretaceous marine sedimentary rocks that indicated latitudinal offset of tectonostratigraphic terranes identified along the California coast. The paleomagnetism of these terranes showed *c.* 1000 km or *c.* 10° of northward latitudinal transport since the Cretaceous, and this amount of offset suspiciously resulted from inclinations that were *c.* 10–15° too shallow in marine sedimentary rocks that were fine grained and rich in clay. Laboratory compaction work had revealed that clay was a major cause of inclination flattening as the magnetite particles stuck to clay particles and rotated into the horizontal with them as compaction proceeded. The magnitude of inclination shallowing in the laboratory was always *c.* 10–15°, so it looked quite plausible that the Cretaceous sedimentary rocks from western North America had suffered from inclination

shallowing rather than from northward tectonic transport (Fig. 5.4).

We continued our study of California tectonstratigraphic terranes with Tan & Kodama's (1998) study of the Point Loma Formation and the Ladd and Williams Formation from southern California. The work followed the same general strategy of the Kodama and Davi study with disaggregation of the marine sedimentary rocks, laboratory compaction of the slurries created from the sedimentary rock material and ARM anisotropy and paleomagnetic measurements; this time however the results indicated that all of the anomalously shallow inclinations could be explained by compaction-induced inclination shallowing. The magnitude of the inclination flattening ranged over 12–20° ($f = 0.56$–0.65).

We finished up our California tectonostratigraphic terrane studies in Baja California by collecting Cretaceous rocks from the Vizcaino Peninsula (Fig. 5.5), where terranes outboard of the main tectonostrati-

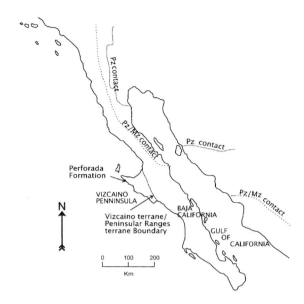

Fig. 5.5 Location of the Perforado Formation on the Vizcaino Peninsula of Baja California (modified from Vaughn *et al.* 2005). The Vizcaino Peninsula is separated from Baja California by a possible tectonostratigraphic terrane boundary. The Pz (Paleozoic) and Pz/Mz (Paleozoic/Mesozoic) contacts show the geologically determined extent of paleolatitudinal offset for Baja California. Reprinted from *Earth and Planetary Science Letters*, 232, J Vaughn, KP Kodama and DP Smith, Correction of inclination shallowing and its tectonic implications: The Cretaceous Perforada Formation, Baja California, 71–82, copyright 2005, with permission from Elsevier.

graphic terranes of California were identified. Since we had shown that the main Peninsular Ranges terrane of western North America had been in place with Tan & Kodama's (1998) study, the question was now: could terranes outboard of the Peninsular Ranges terrane, located in the Vizcaino Peninsula, show latitudinal transport even after compaction had been accounted for? The Perforada Formation showed beautifully the effects of different amounts of compaction on the paleomagnetic inclination (Vaughn *et al.* 2005). Rocks that had strong oblate compaction fabrics for ARM anisotropy and for AMS had the shallowest inclinations. Rocks with very weak magnetic fabrics had inclinations about 10° steeper (Fig. 5.6). Furthermore, when the rocks with strong fabrics had their inclination corrected by the anisotropy technique, the corrected inclination agreed nicely with the uncorrected inclination from the rocks with little or no fabric. This was compel-

ling evidence supporting the anisotropy–inclination correction and the validity of the anisotropy correction approach.

TEST OF THE ANISOTROPY–INCLINATION CORRECTION

The best test of the anisotropy–inclination correction for magnetite-bearing rocks is the study of the Paleocene Nacimiento Formation (Kodama 1997). In this study, the reality of rock magnetically-caused inclination shallowing was well established by a 104 site magnetostratigraphic study (Butler & Taylor 1978) that had a tight (3.0°) 95% confidence limit. Tectonic paleolatitudinal transport of the site could not be used to explain its 7–8° anomalously shallow inclination since the rocks were located in the Colorado Plateau in the Early Tertiary, deep in North America's interior. Furthermore, the cratonic paleomagnetic pole for North America for this period was well determined by igneous rocks (Diehl *et al.* 1983; Besse & Courtillot 2002) that would not be affected by inclination shallowing. The age of the Nacimiento Formation was also well constrained by the magnetostratigraphy measured for the unit, so the comparison to the Paleocene cratonic paleopole wasn't affected by age inaccuracies.

Paleomagnetic and ARM anisotropy measurements were made on the mudstones and siltstones of the Nacimiento Formation and the individual particle anisotropy was determined by three independent techniques: laboratory compaction experiments, direct measurement of a magnetic separate and examination of the shape of the magnetic grains by SEM. All three approaches gave an individual particle anisotropy of *a c.* 2.2–2.8. The inclination of each individual sample was corrected by its anisotropy, and the corrected directions (using the sample declination) were averaged by Fisher statistics (Fisher 1953) as was typically done in paleomagnetic studies. The resulting anisotropy-based inclination-shallowing correction perfectly removed the 7–8° of observed shallowing to bring the Nacimiento direction into agreement with the Paleocene cratonic field for North America (Fig. 5.7). This study clearly showed the power and accuracy of the anisotropy correction. It also supported one of the assumptions used in inclination corrections that the ARM anisotropy should be applied only to the magnetic grains that carried the demagnetized remanence. This can be done relatively easily with an ARM applied to relatively low-coercivity magnetite

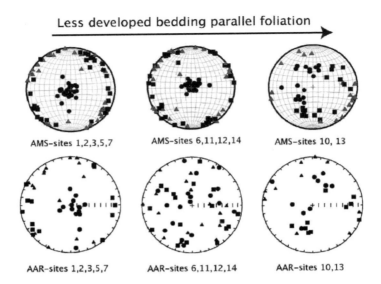

Less developed bedding parallel foliation

AMS–sites 1,2,3,5,7   AMS–sites 6,11,12,14   AMS–sites 10, 13

AAR–sites 1,2,3,5,7   AAR–sites 6,11,12,14   AAR–sites 10,13

Fig. 5.6 Stereonets showing different magnetic fabrics for the Perforada Formation (Vaughn *et al.* 2005). As both AMS and AAR (ARM anisotropy) decrease, the paleomagnetic inclination of the rocks becomes steeper, supporting the use of magnetic anisotropy to detect and correct for inclination shallowing. Reprinted from *Earth and Planetary Science Letters*, 232, J Vaughn, KP Kodama and DP Smith, Correction of inclination shallowing and its tectonic implications: The Cretaceous Perforada Formation, Baja California, 71–82, copyright 2005, with permission from Elsevier.

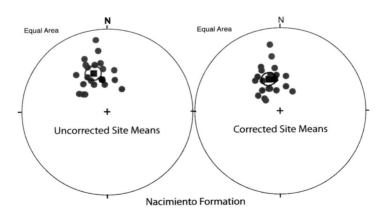

Fig. 5.7 Site mean directions before (left) and after (right) individual sample-by-sample magnetic anisotropy correction of inclination shallowing (Kodama 1997). The mean of the site means (square with 95% confidence circle) moves from shallower than the expected Paleocene direction for North America, shown by gray hexagon (based on Diehl *et al.* 1983), to nearly exactly in agreement with it after correction. KP Kodama, A successful rock magnetic technique for correcting paleomagnetic inclination shallowing: Case study of the Nacimiento Formation, New Mexico, *Journal of Geophysical Research*, 102, B3, 5193–5205, 1997. Figure 14, page 5202. Copyright 1997 American Geophysical Union. Reproduced by kind permission of American Geophysical Union. (See Colour Plate 6)

particles, but becomes more difficult when an anisotropy correction is applied to high-coercivity hematite-bearing red bed sedimentary rocks.

It is important to stress that remanence anisotropy is far superior to susceptibility anisotropy (AMS) for an inclination-shallowing correction. The magnetic response of a rock to an applied magnetic field (susceptibility) is the result of all the magnetic minerals in a rock: clays, ferromagnesian silicates and ferromagnetic minerals such as magnetite and hematite. Even calcite in a carbonate rock can contribute to an AMS. The accuracy of an inclination correction relies on measuring the anisotropy of only the ferromagnetic minerals in a rock, particularly the same subpopulation of ferromagnetic grains that carry the demagnetized remanence.

ANISOTROPY–INCLINATION CORRECTIONS: THE NEXT GENERATION

A survey of the literature shows that by early 2011 ten studies had been reported that used ARM to measure the magnetic anisotropy of a rock in order to make an inclination-shallowing correction (Table 5.1). The magnetic mineralogy of these rocks was typically magnetite; in one case (the Nacimiento Formation) the magnetic mineralogy was a titano-hematite that had a low coercivity like magnetite and therefore ARM could be efficiently applied. Six of these studies (Collombat et al. 1993; Hodych & Bijaksana 1993; Kodama & Davi 1995; Kodama 1997; Tan & Kodama 1998; Vaughn et al. 2005) have already been discussed.

The Kodama & Ward (2001) study listed in Table 5.1 did not make any actual inclination corrections, but used previously corrected rock units to establish the northern limits of rudists in the Cretaceous along western North America in order to constrain the southern limit of British Columbian tectonostratigraphic terranes. The Baja BC controversy (Cowan et al. 1997) about the shallow paleomagnetic inclinations in rocks from coastal British Columbia that put that part of the world near to Baja California in the Cretaceous needed to be addressed by inclination-shallowing corrections. The Kodama & Ward (2001) study attempted to do that by recognizing that because Baja BC rocks did not contain rudist fossils, Baja BC could be no further south than the northern recognized limit of rudist corals in the Cretaceous. This

forced Baja BC to reach no further south than northern California (40°N) based on the inclination corrected locations of rudist fossils. Kim & Kodama (2004) then made a direct inclination-shallowing correction to Baja BC rocks exposed near Vancouver Island. The Nanaimo Group rocks yielded a corrected inclination that beautifully placed the southern end of Baja BC at about 41°N, in agreement with the rudist distribution from corrected paleolatitudes. Krijgsman & Tauxe (2006) also worked on the inclination shallowing of Baja BC rocks, but approached it using the EI technique. Interestingly, Krijgsman and Tauxe came up with the same size correction as Kim and Kodama (Kim and Kodama: $f = 0.7$; Krijgsman and Tauxe: $f = 0.68$), but arrived at different tectonic conclusions based on their correction. Part of the reason for this is that they also included data from the original study (Ward et al. 1997) that had initial inclinations much lower than those obtained by Kim & Kodama (2004) and Enkin et al. (2001) in subsequent studies of the rocks.

The remaining two ARM anisotropy studies in Table 5.1 pushed back the inclination-shallowing correction into Carboniferous age rocks in order to help resolve disagreements about the paleogeographic reconstructions of the continents in this time period. This work, and Tamaki et al.'s (2008) study that used an IRM applied at 45° to bedding (Hodych & Buchan 1994) will be discussed when we consider the inclination-shallowing correction for hematite-bearing rocks.

THEORETICAL BASIS OF THE ANISOTROPY–INCLINATION CORRECTION FOR HEMATITE-BEARING ROCKS

The theoretical basis of the anisotropy–inclination correction provided by Jackson et al. (1991) assumes that the magnetization of the magnetic particles in a rock lies parallel to the long axis of the particle (the 'easy' axis). This is certainly true for a magnetic mineral such as magnetite whose magnetization is governed by the shape of the magnetic particles. Hematite, the magnetic mineral typically carrying the paleomagnetism of red sedimentary rocks (i.e. red beds), has a magnetization controlled not by the shape of the grains but by the internal crystallographic structure of the hexagonal hematite crystal. In the case of hematite, the magnetization of the mineral is constrained to lie in the basal crystallographic plane. The

Table 5.1 Anisotropy-based inclination-shallowing correction

Formation and age	Locality	Magnetic mineralogy	Magnetic anisotropy	Flattening factor, f	Reference
DSDP cores, Cretaceous	Pacific Ocean	Magnetite	ARM	0.53	Hodych et al. (1999)
Marine cores, Recent	Atlantic Ocean	Magnetite	ARM	–	Collombat et al. (1993)
Pigeon Point Formation, Cretaceous	California	Magnetite	ARM	0.7	Kodama & Davi (1995)
Nacimiento Formation, Paleocene	New Mexico	Titano-hematite	ARM	0.84	Kodama (1997)
Point Loma and Ladd Formation	S. California	Magnetite	ARM	0.56 (PL) 0.65 (Ld)	Tan & Kodama (1998)
Cretaceous rudist-bearing rocks	California	Magnetite	ARM	–	Kodama & Ward (2001)
Nanaimo Group, Cretaceous	Vancouver Island, Brit Col.	Magnetite	ARM	0.7	Kim & Kodama (2004)
Perforada Formation, Cretaceous	Baja California, Mexico	Magnetite	ARM	0.6	Vaughn et al. (2005)
Glenshaw Formation Carboniferous	Pennsylvania	Magnetite	ARM	0.65	Kodama (2009)
Deer Lake Formation, Carboniferous	Newfoundland Canada	Magnetite	ARM	0.54	Bilardello & Kodama (2010b)
Mauch Chunk Formation, Carboniferous	Pennsylvania	Hematite	AMS-chemical	0.22	Tan & Kodama (2002)
Kapusaliang Formation, Cretaceous	Western China	Hematite	AMS-chemical	0.3	Tan et al. (2003)
Shepody and Maringouin Formation, Carboniferous	Nova Scotia	Hematite	IRM-high field	0.64 (Shep) 0.83 (Mar)	Bilardello & Kodama (2009b)
Mauch Chunk Formation, Carboniferous	Pennsylvania	Hematite	IRM-high field	0.49	Bilardello & Kodama (2010a)
Cretaceous red beds	SE China	Hematite	IRM-45°	0.82	Wang & Yang (2007)
Yezo Supergroup, Cretaceous	Japan	Magnetite	IRM-45°	0.71	Tamaki et al. (2008)
Elatina and Nucaleena Formation, Neoproterozoic	Australia	Hematite	IRM-parallel and perpendicular to bedding	0.92 (E arenites) 0.97 (E rhythmites) 0.98 (N)	Schmidt et al. (2009)

crystallography of the hematite however controls the shape of the nano-particles of hematite to be plate-like and the basal plane magnetization then typically lies in the plane of the hematite plate-like crystals. Tan & Kodama (2003) therefore modified the Jackson et al. (1991) approach for an orientation distribution of magnetic grains that are disk-shaped with their magnetizations lying in the plane of the disks. This resulted

in a new expression for f, the flattening in the tan $I_c = f$ tan I_0 equation used for the anisotropy correction:

$$f = \frac{(2a+1)K_{\min}-1}{(2a+1)K_{\max}-1}$$

where a is the individual particle anisotropy, which for hematite is typically c. 1.4–1.45 (Kodama 2009), and

K_{min} and K_{max} are the minimum and maximum principal anisotropies for the remanent anisotropy used for the inclination correction of hematite-bearing rocks, i.e. the maximum and minimum axes of the anisotropy ellipsoid. The remanence anisotropy is a bit trickier to measure for hematite-bearing rocks because hematite has such a high coercivity. ARM will not work because most laboratories do not have equipment that can apply an ARM to hematite-bearing rocks. We will see that different approaches have been used to measure the anisotropy of the hematite grains carrying the demagnetized remanence of red sedimentary rocks.

It is worth noting that the all-important individual particle remanence anisotropy can vary widely for magnetite because it is controlled by shape; for hematite however, since the individual particle anisotropy is controlled by crystallography, it tends to be fairly constrained in magnitude. This is a benefit to inclination-shallowing corrections for hematite-bearing rocks since a fairly good correction can result from using a value close to 1.4–1.45 and the difficult procedures for measuring the hematite individual particle anisotropy can be avoided (Kodama 2009).

RED BED INCLINATION SHALLOWING

Making inclination corrections to magnetite-bearing marine sedimentary rocks is fairly non-controversial if the distinct possibility of inclination shallowing is accepted; because magnetite is recognized as a primary magnetic mineral in sedimentary rocks, it could be affected by post-depositional compaction. Inclination corrections of red sedimentary rocks, i.e. red beds, are controversial simply because there's no consensus within the paleomagnetic community about whether the paleomagnetism of red beds is a primary depositional remanence or a secondary chemical remanent magnetization acquired when magnetic minerals are formed post-depositionally. This controversy has been denoted the 'red bed controversy' (see Butler 1992 for more details). Despite the lack of a clear consensus about the magnetization mechanism of red beds, there is one simple reason to pursue an inclination correction for red beds. They have a strong, stable paleomagnetism, are abundant in the geologic record and are a common target of paleomagnetic studies. Furthermore, anisotropy measurements made for inclination corrections can shed some important light on the timing and manner of red bed magnetization acquisi-

Fig. 5.8 Red dots show paleopoles based on red beds for the Mesozoic and Paleozoic. Open squares show Van der Voo's (1990) mean paleopoles for the lower Tertiary (Tl), upper Cretaceous (Ku), upper Jurassic (Ju), lower Jurassic (Jl), upper Triassic (Tru), lower Triassic (Trl), upper Permian (Pu), lower Permian (Pl), upper Carboniferous (Cu), lower Carboniferous (Cl), upper Devonian (Du), lower Devonian (Dl), upper Silurian/lower Devonian (Su/Dl) and middle Ordovician (Om). The more modern APWP of Besse & Courtillot (2002) only extends back to 200 Ma, or between the Tru and Jl mean paleopoles. North America's APWP for earlier times relies heavily on results from red beds. (See Colour Plate 7)

tion and help answer the question about whether the red bed remanence is primary.

The importance of red bed paleomagnetic data is demonstrated by the dominance of red bed paleomagnetic poles in North America's Mid-Paleozoic–Early Mesozoic apparent polar wander path (APWP). When the mean North American paleopoles are compared to the red bed paleopoles (Fig. 5.8), it is apparent that the position of the mean paleopoles is controlled to large extent by the red bed data from the Upper Devonian–Lower Jurassic.

The importance of red beds for determining a continent's ancient APWP is particularly acute because the preferred APWPs, those constructed synthetically for each continent from an averaging of the global paleomagnetic pole database (Besse & Courtillot 2002), only extend back in time 200 million years. Before that time, paleogeographic reconstructions of the continents rely heavily on averaging of individual paleomagnetic poles in different time periods and, as we have shown, many of those paleopoles are derived from red beds. In Besse & Courtillot's (2002) construction of synthetic APWPs for the continents, errors that may occur in one type of lithology (i.e. sedimentary inclination shallowing) are probably minimized by averaging sedimentary and igneous paleopoles. However, any errors are still buried in the data and probably cause a small bias in the synthetic paths, particularly when they're dominated by sedimentary rock results.

The validity of an inclination-shallowing correction for red beds depends on whether the red beds acquired their paleomagnetism at or soon after deposition. Some paleomagnetic studies have suggested that hematite-bearing red sedimentary rocks carry a depositional remanence. If this is the case, red bed remanence probably acquires a magnetization significantly shallower than the geomagnetic field. The natural and laboratory re-deposition of the Siwalik Group hematite-bearing sediments studied by Tauxe & Kent (1984), already discussed in Chapter 2, showed approximately 25° of syn-depositional inclination shallowing. A red bed depositional remanence would also be susceptible to burial compaction. In fact, the inclination of hematite-bearing red beds may be affected more than magnetite-bearing sediments because a red bed depositional remanence could be flattened by both depositional and compaction processes. The reason for the possibility of greater shallowing in red beds is that, in contrast to magnetite particles, relatively large hematite grains (5–20 μm) are stable uniformly magnetized grains (called single domain by paleomagnetists) that carry the red bed's demagnetized remanence. These large grains will be the most susceptible to gravitational forces during deposition and, because hematite has an intrinsic magnetization 200 times smaller than magnetite, will be less likely to be realigned parallel to the geomagnetic field by post-depositional processes.

A chemical origin for red bed remanence, acquired by the growth of secondary hematite grains, has been the more generally accepted magnetization mechanism for red beds. In this case, red bed paleomagnetism should be free of the errors inherent to a depositional magnetization but then the timing of the remanence acquisition becomes an important consideration. Some early workers had suggested that red bed remanence is a chemical remanence acquired millions of years after deposition (Walker *et al.* 1981; Larson *et al.* 1982).

Thus, red bed remanence would not be useful as a record of the geomagnetic field at or near the time of the rock's deposition. However, subsequent extraction of detailed magnetostratigraphies from red bed units shows that even if red bed remanence is a secondary CRM, it must be acquired soon after deposition (thousands or maybe tens of thousands of years). Evidence from one study shows that a red bed chemical remanence was acquired before burial by only 1 m of overburden (Liebes & Shive 1982). These observations indicate that red bed remanence (even if a secondary chemical magnetization) can be acquired soon after deposition and could still suffer from inclination shallowing caused by burial compaction. Even a late-acquired post-compaction chemical magnetization could suffer from inclination shallowing if hematite is formed from precursor minerals with depositional or compaction fabrics.

A compilation of magnetic fabrics measured in red beds is consistent with red sedimentary rocks being dominated by a primary depositional remanence or an early compaction-affected chemical remanence (Fig. 5.9).

These data are particularly compelling because they come from Paleozoic and Mesozoic formations that are considered 'classic' red beds. They all show the typical magnetic fabric, both for anisotropy of magnetic susceptibility (AMS) and anisotropy of remanence (anisotropy of isothermal remanence or AIR) that occurs during deposition and compaction. The oblate fabrics have minimum principal axes oriented perpendicular to the bedding plane and the maximum and intermediate principal axes are distributed in the bedding plane.

THE RED BED MAGNETIC ANISOTROPY–INCLINATION CORRECTION

Because of the high coercivity of hematite, paleomagnetists have tried different methods for measuring the remanence anisotropy of the hematite grains. Tan *et al.*'s (2003) study of the shallow inclinations in central Asian Cretaceous red beds illustrates an approach that uses both chemical demagnetization and the measurement of susceptibility anisotropy (AMS). In this technique, chemical demagnetization is used to isolate the ancient magnetization in the rocks. In chemical demagnetization the rock samples are soaked for progressively longer time periods in concen-

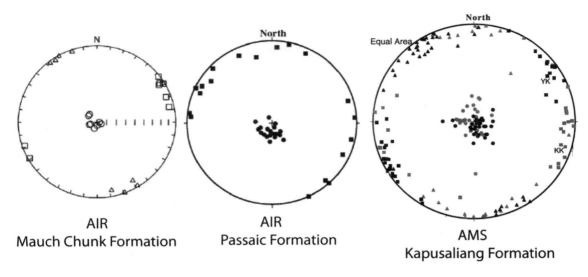

Fig. 5.9 Anisotropy of isothermal remanence (AIR) for the Mauch Chunk and Passaic Formations and chemically demagnetized anisotropy of magnetic susceptibility (AMS) for the Kapusaliang Formation. These magnetic fabrics are observed in typical red bed formations plotted in stratigraphic coordinates. Note that the magnetic fabrics are characteristic of deposition and/or compaction with minimum axes (circles) perpendicular to bedding, indicating that the hematite carrying the remanence has depositional/compactional fabrics. Squares: maximum axes; triangles: intermediate axes. All points lower hemisphere.

trated acid. As the acid soaks into the samples, first the smallest then eventually the largest hematite particles are dissolved. Rinsing and measuring the samples periodically during the acid leaching monitors the removal of the magnetization. The underlying assumption of the chemical demagnetization of red beds is that the smallest hematite grains are the pigmentary grains and are secondary, while the largest hematite grains are detrital and therefore primary. AMS is also measured at each chemical demagnetization step. Tan *et al.* (2003) were interested in determining the AMS removed at the same chemical demagnetization steps that isolated the ancient magnetization direction. This was done by fitting a second-rank AMS tensor at each chemical demagnetization step and determining, by subtraction, the second-rank AMS tensor removed between the same chemical demagnetization steps that isolated the ancient magnetization direction in the rock samples. In this way, the AMS of the hematite grains that carries the ancient remanence can be used to correct the inclination. This approach relies on the ferromagnetic hematite grains dominating the AMS and on the chemical demagnetization isolating the same magnetic direction as thermal demagnetization,

the technique typically used to identify the ancient magnetization in red beds.

These assumptions were fulfilled beautifully with the Kapusaliang Formation from China. Determining the individual particle anisotropy for susceptibility, which has a different value than the remanence particle anisotropy for the same grains, was however more difficult. It could not be measured directly, but could only be derived by finding the best fit between the corrected inclinations and the theoretical correction curves. The value obtained was later compared to the directly measured remanence particle anisotropy for the Mauch Chunk red beds and found to be in good agreement. An individual particle susceptibility anisotropy of 1.06 led to a 32.5° decrease in inclination for an $f = 0.30$, a very large amount of shallowing.

The Kapusaliang results were significant because they were the first results to show that the anomalously shallow paleomagnetic inclinations observed in Cretaceous and Early Tertiary rocks from central Asia were not due to standing non-dipole fields or unrecognized continental-scale shear zones, as had been previously proposed, but due to inclination shallowing in red beds. The shallow inclinations in

Central Asia had presented a geological problem for paleomagnetic data because there was no evidence of over 1000 km of intra-continental shortening to the north.

Another approach that was used to measure the fabric of red bed hematite grains employed an isothermal remanent magnetization (IRM) and many high-temperature heatings. The IRM was applied in high fields (c. 1 T) in the same orientations used in the ARM anisotropy experiments. The IRM application was followed by a thermal demagnetization at about 630°C for each orientation to measure the IRM carried by the same hematite grains that carried the ancient magnetization isolated by standard thermal demagnetization of the samples. To apply an IRM, the rock sample is simply exposed to a DC magnetic field. It is a different way of applying a laboratory remanence and can reach higher fields than an ARM application, thus allowing activation of very high-coercivity hematite grains. After measurement of the thermally demagnetized IRM, the samples were further thermally demagnetized at 690–700° to totally remove the IRM applied at each orientation before moving on to the next orientation. This approach was used in the Tan et al. (2007) study of the Triassic Passaic Formation red beds of New Jersey and requires many high-temperature heatings that can cause chemical and mineralogical alteration of the hematite grains. Despite this, the study of the Passaic Formation red beds gave very reasonable results and provided an important comparison to the EI correction technique (see next section).

Bilardello & Kodama (2009a) developed one more way to measure the remanence anisotropy of the high-coercivity hematite grains in red beds. Their approach was initially proposed by Kodama & Dekkers (2004), but Bilardello & Kodama (2009a) provided a full development and test of the procedure. Kodama and Dekkers applied IRMs in very high fields (13 T) at a special strong magnet facility in Nijmegen, Holland so that no demagnetization was necessary between orientations; the magnetization of the grains was totally reset at each orientation. Obviously, this approach was not practical for most paleomagnetic studies since only four of these strong magnet facilities exist in the world. Bilardello & Kodama (2009a) used very small samples, so that a standard impulse magnetizer with a high field coil could be used to apply 5 T IRMs to samples in the different orientations used to determine the anisotropy of isothermal remanence (AIR). Using an alternating field demagnetization, Bilardello and Kodama then demagnetized the samples at 100 mT to remove the contribution of any magnetite that might be in the samples. Rock magnetic measurements had shown the presence of magnetite in addition to the hematite typically found in red beds (Kodama & Dekkers 2004; Bilardello & Kodama 2009a; Kodama 2009). Finally, Bilardello and Kodama used low-temperature thermal demagnetization at 120°C to remove the contributions due to goethite, a secondary magnetic mineral that can be common in red beds. The high fields used in these experiments to totally reset the IRM at each orientation would also activate any goethite present.

The technique presented in Bilardello & Kodama (2009a) was successfully used to isolate the remanence anisotropy for inclination corrections of the Carboniferous red beds of the Maritime Provinces of Canada (Bilardello & Kodama 2009b). In this study, the inclinations of the Shepody Formation and Maringouin Formation were corrected (Shepody: $f = 0.64$; Maringouin: $f = 0.83$) to provide an improved Carboniferous paleomagnetic pole for North America. Interestingly, the corrected paleopoles for the nearly coeval Shepody and Maringouin Formations fall almost exactly on top of each other while the uncorrected poles were several degrees apart. The Shepody and Maringouin results could be compared to a magnetite correction of the Glenshaw Formation of western Pennsylvania and a hematite correction of the Mauch Chunk Formation red beds of eastern Pennsylvania (Kodama 2009; Bilardello & Kodama 2010a). The corrected and coeval paleopoles were found to be in good agreement. The Mauch Chunk correction reported by Bilardello & Kodama (2010a) used the high field AIR technique of Bilardello & Kodama (2009a) and had less of a correction ($f = 0.49$) than the chemical demagnetization correction of the Mauch Chunk ($f = 0.22$) by Tan & Kodama (2002). The high field AIR correction seems to be the best way to make an anisotropy correction to hematite-bearing red beds because it involves less potential mineralogical alteration of the magnetic grains by either acid leaching or repeated high temperature heating.

A SIMPLIFIED ANISOTROPY CORRECTION

One of the most difficult aspects of the anisotropy–inclination shallowing correction for red beds is the determination of the individual particle anisotropy, particularly if it is for susceptibility when the chemically demagnetized AMS is used to make a correction. The papers mentioned here all used a curve-fitting technique of the anisotropy data to the theoretical correction curves to find the best-fit individual particle anisotropy. This approach has been shown to be effective in Kodama's (2009) study of the Carboniferous Glenshaw formation, but extraction and direct measurement of the remanence anisotropy of individual hematite grains suggest that they have similar values in the range c. 1.4–1.45, as mentioned previously. Furthermore, Kodama (2009) found that the anisotropy correction could be further simplified by using an average bulk anisotropy for a rock unit to correct the average paleomagnetic direction (formation mean direction) from the unit. This would suggest that previously reported paleomagnetic results could be corrected by the average remanence anisotropy measurements of some strategically collected samples from a unit, and would allow quick propagation of corrected directions and paleopoles in the literature. The individual particle remanence anisotropy could be determined by curve-fitting techniques or by more laborious direct measurement of extracted magnetic particles.

An inclination correction of the mean paleomagnetic direction, rather than a sample-by-sample or site-by-site correction as is usually done, has been shown to be the same as a sample-by-sample correction if the procedure outlined by Kim & Kodama (2004) for the magnetite-bearing rocks of the Northumberland Formation is followed. In this approach, the mean direction vector is multiplied by the inverse of the DRM tensor in order to obtain the corrected mean direction:

$$H_i = \mathbf{DRM}^{-1} \mathrm{NRM}_i$$

where H_i are the components of the corrected mean direction, NRM_i are the components of the uncorrected but demagnetized mean direction and $\mathbf{DRM}^{-1}$ is the inverse of the mean DRM tensor. The idea of the DRM tensor derives from Jackson et al. (1991) and is related to the orientation distribution of the magnetic particle easy axes in the rock. The mean DRM tensor is obtained from the anisotropies of individual samples and averaging them, tensor element by tensor element. The same coordinate system should be carefully maintained for all the samples during the averaging and application of the above equation. Kodama (2009) used only 23 samples from the Glendale Formation and

obtained remarkably similar results to a sample-by-sample correction. One of the underlying assumptions of the anisotropy correction is that, by using the maximum and minimum remanence anisotropy axes in the hematite and magnetite theoretical equations, the minimum axis is always perpendicular to bedding and the maximum axis is always parallel to the declination. This, obviously, is rarely true. The use of the DRM tensor avoids making this assumption, accurately using the orientation distribution of the magnetic particles to correct the paleomagnetic direction.

The DRM tensor can be derived directly from remanence anisotropy measurements and the individual grain anisotropy. As Kim & Kodama (2004) indicate, that relationship for magnetite is:

$$\mathrm{DRM}_i = \frac{K_i(a+2)-1}{a-1}$$

where DRM_i are the principal components of the DRM tensor and K_i are the principal components of the remanence anisotropy tensor, usually ARM for magnetite. The individual particle anisotropy is a. For hematite-bearing rocks, the relationship between the DRM tensor and the remanence anisotropy tensor is:

$$\mathrm{DRM}_i = \frac{1 - K_i(1+2a)}{1-a}$$

where K_i, the principal axes of the remanence anisotropy tensor, can be derived from high field AIR. A particle anisotropy of c. 1.4 is likely to give good results.

PROPAGATING ERRORS IN THE ANISOTROPY–INCLINATION CORRECTION

The picture is not complete unless the errors associated with an anisotropy-based inclination correction are propagated through the method. Until now, an estimate of the error for anisotropy–inclination corrections comes only from Fisher statistics (Fisher 1953) applied to the corrected sample directions or corrected site mean directions. The errors that come from estimating the flattening factor f, which is based on the anisotropy measurements and the determination of the individual particle anisotropy factor a, have not been included. Bilardello et al. (2011) have propagated the errors in f using bootstrap statistics and found, in a case study of the Shepody Formation inclination correction (Bilardello & Kodama 2010b) that there was a 15% uncertainty in determining f. This scale uncertainty led only to a 0.31° increase in the 95% cone of confidence for the mean corrected direction and a steepening of the corrected mean inclination by 0.32°. Therefore, although propagating uncertainties in f should be calculated for a complete inclination correction, it appears that error propagation has a very small effect on the mean corrected direction inclination and cone of confidence.

THE 45° IRM CORRECTION

Tamaki *et al.* (2008) and Wang & Yang (2007) are two of the three remaining inclination-shallowing correction studies (Table 5.1) that have not yet been discussed. Both Tamaki *et al.* and Wang and Yang used Hodych & Buchan's (1994) approach of applying an IRM at 45° to bedding to see if the rock had a strong enough remanence anisotropy to deflect its remanence from the geomagnetic field inclination during the rock's acquisition of its paleomagnetism. Tamaki *et al.* studied magnetite-bearing rocks and Wang and Yang studied hematite-bearing rocks. Hodych and Buchan used the technique to study the Silurian Springdale Group red beds of Newfoundland. They saw little deflection of the IRM they applied and concluded that inclination shallowing had not occurred. Stamatakos *et al.* (1994) commented on this study, indicating that the uncorrected remanence of the red beds was much shallower than the magnetization of nearby volcanics of the same age and did not give a reasonable paleogeographic interpretation for the Silurian. Smethurst & McEnroe (2003) restudied the paleomagnetism of the Springdale volcanics and red beds showing that, with more rigorous demagnetization, the difference in inclination between the volcanics and red beds disappeared, thus vindicating the conclusion of Hodych & Buchan (1994) that inclination shallowing had not affected these red beds.

Tan *et al.*'s (2002) re-deposition work with red bed material from China was designed to test Hodych and Buchan's 45° IRM approach for detecting inclination shallowing. The test is much easier to conduct than the time- and labor-intensive measurement of a rock's complete remanence anisotropy with many applications of laboratory remanences in different orientations. It would also allow a typical paleomagnetism laboratory to easily check the anisotropy of high-

coercivity red beds. Tan *et al.* (2002) re-deposited red bed sediment, which spontaneously separated into fine-grained and coarse-grained fractions (see Chapter 4), and observed significant compaction shallowing for the fine-grained fraction and little compaction shallowing for the coarse-grained fraction. The application of the IRM at 45° (Hodych and Buchan's prescription) and 60° (the initial inclination during Tan *et al.*'s redeposition) accurately indicated significant anisotropy and deflection for the IRM applied to the fine-grained sediments. The amount of deflection for the 60° IRM was similar to the amount of compaction shallowing. Similarly, the coarse-grained fraction showed little deflection of the IRMs.

The results of this experiment are good news for the Hodych and Buchan approach. Unfortunately however, for a separate set of re-deposition experiments in which significant syn-depositional inclination shallowing of the coarse-grained fraction occurred, the 45° IRM was unable to detect any anisotropy and shallowing even though it existed. Tan *et al.* (2002) further tested the 45° IRM approach on two units corrected by complete measurement of the anisotropy tensor, the Mauch Chunk Formation and the Passaic Formation. The 45° IRM technique worked well for the Mauch Chunk, but did not detect shallowing for the Passaic.

We will show in the following section that the Passaic result agrees nicely with an EI correction for the same unit, so it is particularly troubling that the 45° IRM method doesn't detect any shallowing for the Passaic Formation. The 45° IRM approach appears to be particularly sensitive to magnetic particle grain size and can give false negative results (i.e. no shallowing when there is some). If the technique detects shallowing it is probably correct, but if it indicates no shallowing this negative result should be checked by the measurement of the complete anisotropy tensor.

Schmidt *et al.*'s (2009) inclination shallowing study of the Snowball Earth deposits in Australia (Table 5.1) and the Eltanin and Nucaleena formations is similar in that application of high field IRMs (12 T) parallel and perpendicular to bedding did not show significant anisotropy or shallowing. Although this study was more complete than a simple 45° IRM application, it perhaps should be further tested with the measurement of a complete anisotropy tensor. Schmidt *et al.* (2009) did check their results with application of the EI technique and it yielded slightly steeper corrected inclinations (by 5°) than the three-axis application of an IRM, suggesting that the measurement of a complete tensor might give a different result than simply IRMs applied parallel and perpendicular to bedding.

ELONGATION–INCLINATION CORRECTION TECHNIQUE

Subsequent to the introduction and development of the anisotropy-based inclination-shallowing correction, Tauxe & Kent (2004) were inspired to approach shallow inclinations in sedimentary rocks from a totally different perspective. They reasoned that inclination shallowing would affect the distribution of site means in sedimentary rocks significantly flattened by either depositional or compaction processes. Tan *et al.* (2003) observed a flattened site mean distribution in their uncorrected site means for the red beds of the Kapusaliang Formation in Central Asia and pointed out that the flattened distribution was probably an indication of inclination shallowing. Tauxe and Kent's elongation–inclination method of correcting for inclination shallowing is a technique for detecting and correcting shallow inclinations totally independently of anisotropy measurements and has gained much popularity in the paleomagnetic community. Table 5.2 shows 14 studies in the literature where the EI method has been used to check for and correct any inclination shallowing detected in sedimentary rocks.

The EI method is both a clever and relatively easy technique that has given essentially the same magnitude inclination correction as the anisotropy-based correction in some important test cases (Tauxe *et al.* 2008).

The EI technique assumes that the Earth's secular variation behavior has remained essentially the same through geological time as it has been for the past 5 million years. This assumption has been tested (Tauxe *et al.* 2008; Tauxe & Kodama 2009) using the geomagnetic secular variation recorded by large igneous provinces from the Precambrian to the Tertiary, and the results show relatively nice agreement with the geomagnetic field model used for the EI inclination correction (Fig. 5.10).

The large igneous province (LIP) data do appear to support the EI field model, particularly at intermediate and high inclinations and very flat inclinations (equatorial fields). So far, not much data are available to check the validity of the model in the low, intermediate inclination range.

The EI technique uses the distribution of the sedimentary rock site means to detect and correct inclination shallowing. The technique assumes that each site is a time-independent spot reading of the secular variation of the geomagnetic field. The field model used by the EI technique (TK03) has a circular distribution of virtual geomagnetic poles (VGPs). A VGP is simply the paleomagnetic north pole that correlates to a paleomagnetic direction somewhere on the globe, assuming that the geomagnetic field is caused by a dipole at the center of the Earth. Because of the non-linear transformation between directions and VGPs, a circular distribution of VGPs transforms to an elongate distribution of paleomagnetic directions. This distribution is elongate along the magnetic meridian (declination) and becomes more elongate as the observation site approaches the equator, where its elongation has the greatest value (*c*. 3). At the north (or south) geomagnetic pole the directions have a circular distribution, just like the VGPs. The elongation is measured by the ratio of the maximum and intermediate eigenvalues of a tensor fit to the distribution. To visualize this, the directional distribution can be fit to a 3D ellipsoid. The shortest axis of the ellipsoid is vertical and the maximum and intermediate axes are horizontal. The eigenvalues are proportional to the axes of the ellipsoid.

The EI correction uses this geomagnetic field behavior to correct for inclination shallowing. If the inclination of a sedimentary rock is flattened by compaction or a syn-depositional process, the directional distribution first becomes circular and then elongate in an east–west direction perpendicular to the magnetic meridian, as the amount of flattening increases. The EI technique assumes that the flattening is caused by the King (1955) $\tan I_o = f \tan I_0$ relationship where f is the same flattening factor introduced in Chapter 4 and used throughout the book to describe inclination shallowing. In the EI technique, no rock magnetic data are used to directly measure the flattening factor f. Different f factors are used to numerically unflatten the site mean distribution. The elongation of the unflattened distribution and its mean inclination are plotted for different flattening factors usually starting with $f = 1$ (no flattening) and then decreasing toward $f = 0$ (complete flattening). Each flattening factor creates a different elongation–inclination pair that plots as a point on the elongation–inclination graph. These points are connected to form a curve. When the curve crosses the elongation–inclination curve of the geomagnetic TK03

Table 5.2 Elongation–inclination (EI) shallowing correction

Formation and age	Locality	Magnetic mineralogy	Flattening factor, *f*	Reference
Miocene sedimentary terrestrial (Spain) and marine (Crete) rocks	Calatayud Basin, Spain and Crete	Hematite (Spain) Magnetite (Crete)	0.68 (Spain) 0.74 (Crete)	Krijgsman & Tauxe (2004)
Oligo-Miocene red beds, Subei and Miocene Siwalik Group red beds	Subei, China and Potwar, Pakistan	Hematite	0.49 (China) 0.77 (Pakistan)	Tauxe (2005)
Late Miocene–Late Pliocene terrestrial strata	Guide Basin, NE Tibet	Hematite	0.6	Yan et al. (2005)
Cretaceous Silverquick Formation and Nanaimo Group marine sedimentary rocks	British Columbia	Magnetite	0.96 (Silverquick) 0.7 (Nanaimo)	Krijgsman & Tauxe (2006)
Late Cretaceous red beds	Jishui and Ganzhou, Jianqi Province, China	Hematite	0.62 (Jishui) 0.73 (Ganzhou)	Wang & Yang (2007)
Triassic Passaic Formation red beds and Paleocene Nacimiento Formation siltstones	Newark Basin, NJ and Colorado Plateau, NM	Hematite-Passaic Titano-hematite-Nacimiento	0.45 (Passaic) 0.84 (Nacimiento)	Tauxe et al. (2008)
Elatina Formation diamictite and Nucaleena Formation cap carbonate	Marinoan Snowball Earth deposits, Australia	Hematite	0.64 (Elatina) 0.74 (Nucaleena)	Schmidt et al. (2009)
Late Cretaceous red beds	Lhasa block, Tibet	Hematite	0.48	Tan et al. (2010)
Early Cretaceous pelagic limestones	Southern Alps, Italy	Magnetite	0.89	Channell et al. (2010)
Jura-Cretaceous limestones	Crimea and Pontides	Magnetite	0.88 (Pontides) 0.94 (Crimea)	Meijers et al. (2010b)
Carboniferous limestones	Donbas Fold Belt, Ukraine	Magnetite	0.65	Meijers et al. (2010a)
Mid-Cretaceous to latest Paleocene limestones	Northern India, Tethyan Himalaya	Magnetite	0.69 (Zongpu) 0.91 (Zongshan)	Dupont-Nivet et al. (2010), Patzelt et al. (1996)
Triassic red beds	Hexi Corridor, China	Hematite	0.94	Liu et al. (2010)
Oligo-Miocene Sespe Formation red beds	California	Hematite	0.73	Hillhouse (2010)

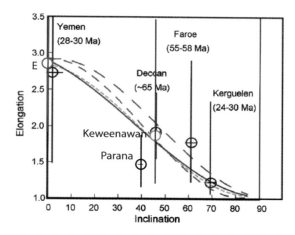

Fig. 5.10 LIP data in support of the TK03 field model used for the EI inclination correction. Red line is the TK03 model, dashed lines are from Constable & Parker's (1988) field model while the dotted line is Quidelleur & Courtillot's (1996) field model. The Parana LIP is Cretaceous (about 120 Ma) and was considered problematic by Tauxe *et al.* (2008). Keweenawan is 1.1 Ga and is from the Northshore basalts of Lake Superior (Tauxe & Kodama 2009), E is the 30 Ma Ethiopian traps of Rochette *et al.* (1988). Plot modified from Tauxe *et al.* (2008). Reprinted from *Physics of the Earth and Planetary Interiors*, 169, L Tauxe, KP Kodama and DV Kent, Testing corrections for paleomagnetic inclination error in sedimentary rocks: A comparative approach, 152–165, copyright 2008, with permission from Elsevier. (See Colour Plate 8)

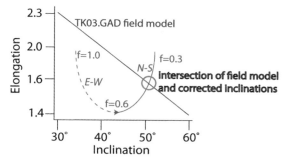

Fig. 5.11 The mechanics of the EI correction technique. The blue curve is the TK03 GAD field model used in the correction. It predicts north–south elongated directional distributions as a function of inclination. The red curve is the directional distribution elongation of a particular set of directions that are being corrected. The directions have an east–west elongation at $f = 1$ (uncorrected) and an inclination of about 35° in this example. As the directions are 'unshallowed', the elongation decreases in an east–west sense according to $\tan I_c = f \tan I_0$ (dashed curve), reaches a minimum (nearly circular) at about 42° and $f = 0.6$ in this example, and then becomes more elongated in a north–south sense (solid curve) until at a particular elongation–inclination pair it crosses the TK03 GAD field model curve (green circle). The intersection is the corrected inclination. In the application of the technique many thousands of corrections are made with bootstrapped data based on the initial directional distribution of directions. This creates many crossing points with the field model. The crossing points give a quantitative measure of the mean crossing point and the 95% confidence interval around the mean. (See Colour Plate 9)

field model, the corrected inclination and flattening factor are determined by the intersection of the two curves (Fig. 5.11). Tauxe & Kent (2004) also use bootstrap statistics to determine confidence limits for the corrected inclination.

In order to obtain a robust correction, many more site means are needed than are typically collected in a paleomagnetic study (>100–150; Tauxe *et al.* 2008), but standard paleomagnetic measurement and demagnetization techniques are all that are required to reduce the data. For this reason, the best EI corrections come from analysis of magnetostratigraphic studies that typically collect many sites. The EI technique was demonstrated, in Tauxe's (2005) introduction of the EI technique, with a correction of central Asian Oligo-Miocene red beds collected near Subei, China that were sampled by Gilder *et al.* (2001) for a magnetostrati-

graphic study. These rocks showed anomalously low inclinations. The study had 222 sites and the inclination was corrected from 43.7° to 63° ($f = 0.4$). Tauxe (2005) also tested the technique with 105 site means from the Siwalik Group of Pakistan (Tauxe & Opdyke 1982) and observed a flattening factor of $f = 0.77$ for the correction ($I_0 = 33.7$, $I_c = 41$; Table 5.2).

One important limitation of the EI technique that is not often recognized is that it is assumed that each site is an independent 'spot' measure in time of geomagnetic field secular variation. Because each sedimentary rock sample will integrate its recording of the field over the time of its deposition, paleomagnetic samples of finite thickness will average some secular variation. Furthermore, unless all the samples from a paleomagnetic site are collected from exactly the same horizon, there will be some averaging of secular variation within a site. The net result will be to smooth some of the features of secular variation and hence reduce the elongation of the site mean distribution recorded by the sedimentary rocks. This could lead to an underestimate of the initial corrected inclination. For example, typical deep-sea sediment accumulation rates are less than 1 cm/kyr; in Chapter 3 deep-sea sediments studied to understand pDRM acquisition varied over the range 1–8 cm/kyr. If a typical deep-sea paleomagnetic sample is a cube of side length 2 cm, it could integrate the field over 250 years to more than 2000 years. Secular variation is averaged out over thousands of years, so for very slow sediment accumulation rates (<1 cm/kyr) the scatter due to geomagnetic secular variation, and needed for an accurate EI correction, would not be observable in a collection of samples. The situation becomes worse if samples at a site are collected over a finite thickness of section. If for instance samples at a site are collected over *c.* 0.3 m of stratigraphic thickness (not unlikely in paleomagnetic studies), the site could average secular variation over 3750 to >30,000 years, more than enough to completely remove the scatter due to secular variation. The EI technique has been applied to many red bed results since its introduction, and for red beds the situation is similar. Sadler (1981) suggests that the sediment accumulation rates for fluvial, continental sediments ranges 1–10 cm/kyr.

A major difference between the EI and anisotropy techniques is that the EI technique does not collect or use any information about the rock magnetic behavior of the sedimentary rocks being studied. The anisotropy-based correction requires that the user measure the remanence anisotropy of at least representative samples, if not all of the samples being studied. The anisotropy technique also requires that the individual particle anisotropy be determined. These data in turn tell the user about the nature of the magnetic fabric, i.e. whether it is primary or secondary and whether a primary or secondary magnetic mineral is carrying the paleomagnetism of the rock. The anisotropy technique requires the user to learn enough about the rock magnetics to help judge the quality of the result. For the EI technique, this information can be assessed with additional rock magnetic measurements not required by the technique.

Tauxe *et al.* (2008) present a comparison of the EI and anisotropy-based correction techniques for two case studies: (1) the red beds of the Triassic Passaic Formation corrected by Tan *et al.* (2007) using the ani-

sotropy technique and (2) the Nacimiento Formation continental sediments corrected by Kodama (1997) using ARM anisotropy (Table 5.2). In each case the anisotropy and EI corrections are statistically indistinguishable from each other. For the Passaic Formation rocks, the anisotropy technique yielded a 10–11° correction while the EI technique gave a 16° correction. The corrected inclinations for the Passaic Formation rocks are 29° for the anisotropy technique and 34° for the EI correction. Tauxe *et al.* (2008) argue that the 200 Ma paleomagnetic pole for North America indicates that the EI correction is closer to reality; however, Tan *et al.* point out that nearby Newark basin volcanics of the same age show a 27° inclination, in better agreement with the anisotropy correction. For the Paleocene Nacimiento Formation of New Mexico, the anisotropy and EI techniques gave remarkably close corrected inclinations (anisotropy: 57°; EI: 56°) that are in agreement with the Paleocene pole for North America that is based on volcanic results (Diehl *et al.* 1983).

The agreement the EI and anisotropy corrections for the red beds of the Passaic Formation and siltstones and claystones of the Nacimiento Formation, as well as with coeval volcanic rock results, bodes well for the robustness and accuracy of each technique. The sampling for the EI corrections was particularly well suited for the technique. The samples from the Passaic Formation were collected from deep cores drilled into the Newark Basin; each data point only sampled the geomagnetic field over the time it took the 25 mm sample to be deposited. The Nacimiento Formation data were from a magnetostratigraphic study with three standard paleomagnetic samples scrupulously collected from a single sedimentary horizon. The average sediment accumulation rate for the Passaic was about 160 m/myr (Kent & Olsen 1997), so a 25 mm sample would average the geomagnetic field over only about 150 years. The average sediment accumulation rate of the Nacimiento Formation was about 60 m/myr (based on the magnetostratigraphy), indicating that a 20 mm sample cube would average the field over about 330 years. Each interval is short enough not to average out secular variation, perfect for the EI correction technique.

One of the pitfalls of using the EI technique is unrecognized vertical axis rotations between sites in a paleomagnetic study. If these rotations are subtle, they could cause an east–west elongation in the site mean distribution that would be misinterpreted to be caused by inclination flattening and yield an overcorrection of the data. This effect is demonstrated in Hillhouse's (2010) study of the Oligo-Miocene Sespe Formation red beds from coastal California (Table 5.2). These Mid-Tertiary rocks show a vertical axis rotation for the area that is consistent with the area's tectonic history. The paleomagnetic results also indicate an anomalously shallow inclination that indicate 11° of post-Mid-Tertiary poleward motion of the area, inconsistent with the amount of movement predicted by motion along the San Andreas fault by a factor of 3. Applied to these rocks, the anisotropy technique increases the inclination from 39° to 47° and into statistical agreement with the expected paleolatitude given by San Andreas motion ($f = 0.73$). In order to have enough sites for the EI correction, additional sites from other studies were added to Hillhouse's dataset increasing N to 158. Individual sample directions had to be used instead of site means, in order to reach a high enough number of samples for the EI correction. Differential rotations between sites were seen in the data and so, to minimize this effect on the correction, each site mean was corrected to a common mean declination of 0 before its individual sample directions were used in the EI technique. Despite this modification of the dataset, the EI technique gave a greater corrected inclination to 51°. The corrected inclination eliminates the necessity of postulating any movement along the San Andreas Fault, so could be interpreted as a slight overcorrection (probably due to vertical axis rotation effects that were not adequately removed).

The studies in Table 5.2 show that the EI correction has been applied to both magnetite-bearing and hematite-bearing sedimentary rocks and resulted in corrected inclinations that give geologically or tectonically reasonable interpretations.

CORRECTING FOR INCLINATION FLATTENING

The experimental and observational results covered in Chapters 4 and 5 show that inclination flattening is a reality in many sedimentary rocks, both marine sediments and terrestrial sediments. It has been observed and corrected in both magnetite-bearing rocks and in hematite-bearing rocks. It is therefore critical that

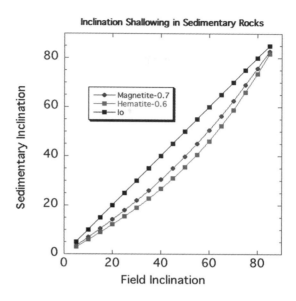

Fig. 5.12 The magnitude of inclination shallowing for typical magnetite and hematite-bearing sedimentary rocks as a function of initial geomagnetic field inclination. (See Colour Plate 10)

accurate use of the paleomagnetic inclination, primarily to determine the paleolatitude of a rock unit, must be checked for inclination shallowing and corrected if it is detected. Either the anisotropy-based technique, as initially envisioned by Jackson *et al.* (1991) and modified for hematite-bearing rocks by Tan & Kodama (2003), or the elongation–inclination technique (Tauxe & Kent 2004; Tauxe 2005) can be used for the detection and correction of inclination flattening, being mindful of the limitations and pitfalls of each correction method. The tabulated flattening factors for magnetite and hematite-bearing rocks in Chapter 4 don't change significantly when the studies in this chapter are added. Magnetite-bearing rocks typically have flattening factors near to 0.7 and hematite-bearing rocks near to 0.6. Fig. 5.12 shows that the maximum inclination error occurs at initial inclinations of *c.* 45–50° and has magnitudes of *c.* 10° and 15° for magnetite- and hematite-bearing sedimentary rocks, respectively.

Post-Depositional Diagenesis and Chemical Remanent Magnetization

The theme of this book is the accuracy and the reliability of paleomagnetic results from sedimentary rocks. In this chapter we'll examine how diagenesis, both early and late, can affect the paleomagnetic signal in sediments and sedimentary rocks. As Tarling points out in the introduction to his book *Palaeomagnetism and Diagenesis in Sediments* (Tarling 1999), historically paleomagnetists were most concerned about the inclination error discovered in the early re-deposition experiments discussed in Chapter 4; diagenetic effects were thought to be minor. This view was changed significantly by Bob Karlin's observation of magnetite dissolution due to early diagenesis in recent hemipelagic marine sediments from the Gulf of California and coastal Oregon (Karlin & Levi 1983; Karlin 1990).

When secondary magnetic minerals first form they are too small to carry a stable remanence; they are superparamagnetic grains with magnetizations that can change direction due to external fields on timescales of only seconds or minutes. As they grow larger and pass through a very narrow size range, they quickly attain stability to external magnetic fields with relaxation times of the order geological timescales (10^9 years). This size is known as the 'blocking volume'

because the grains growing through this size have 'blocked in' a geologically stable magnetization. When they reach this grain size, the grains are magnetized by a chemical (or crystallization) remanent magnetization (CRM). Both Tauxe's (2010) book *Essentials of Paleomagnetism* and Butler's (1992) book *Paleomagnetism: Magnetic Domains to Geologic Terranes* have excellent treatments of the theoretical aspects of CRM acquisition. The pertinent fact we need to keep in mind for our discussion is: when magnetic minerals grow at some time after the deposition of sediments, they will pick up a secondary magnetization at the point in time when their particle grain size grows large enough to become paleomagnetically stable. This grain size is usually of submicron–micron order (e.g. 0.05–1 μm for magnetite; Butler & Banerjee 1975).

What this means for the accuracy and reliability of sedimentary rock paleomagnetism is that the direction of a CRM may be a very accurate record of the geomagnetic field when the magnetic grain grew through its blocking volume, but the magnetization will have an age younger than the depositional age of the rock. How much younger depends on the post-depositional chemical environment of the sediments, i.e. whether early or late diagenesis occurs, and over what time period the

Paleomagnetism of Sedimentary Rocks: Process and Interpretation, First Edition. Kenneth P. Kodama.
© 2012 Kenneth P. Kodama. Published 2012 by Blackwell Publishing Ltd.

chemical changes occur. We will see examples of sediments in which diagenesis creates new magnetic minerals within 1000 years of deposition and other examples in which the diagenetic changes and the subsequent CRM occur millions and tens of millions of years after the rock is formed. The other factor to consider is the duration of time during which the diagenetic changes occur. In some cases the growth of the secondary magnetic minerals occurs over a geologically short period of time (hundreds–thousands of years) and the CRM provides a snapshot of the geomagnetic field at some time after deposition. In other cases, the growth of secondary magnetic minerals occurs over an extended time period and the CRM integrates geomagnetic field behavior over that period. The worst-case scenario is when the diagenetic changes cause a remagnetization of the sedimentary rock at some time long (tens to hundreds of millions of years) after its formation and over an extended period, so that very little useful paleomagnetic information can be extracted from the rock. Paleomagnetists are constantly checking the age of the magnetization of a sedimentary rock by field and laboratory tests as a way of gauging the age, and therefore reliability, of its paleomagnetic direction.

Some of these field and laboratory tests will be discussed in Chapter 9. More detailed and comprehensive treatment can be found in the books of Tauxe (2010) and Butler (1992), which describe techniques such as the fold test, reversal test, conglomerate test, contact test, alternating field demagnetization, thermal demagnetization and principal component analysis. For this book's discussion of CRM and diagenesis, one important fact should be remembered: the magnetic mineralogy can be an important guide as to whether a magnetization is primary (formed at a rock's deposition) or secondary. Magnetite (Fe_3O_4) is almost always a primary depositional magnetic mineral. Specular hematite (Fe_2O_3) in red terrestrial clastic rocks can also be a depositional magnetic mineral. Iron sulfides such as greigite (Fe_3S_4) and usually pyrrhotite ($Fe_{1-x}S$) are secondary magnetic minerals formed by reduction diagenesis in marine and lake sediments. In red beds, submicron-sized pigmentary hematite (which gives the rocks their red color) is secondary. It is typically euhedral in grain morphology. Magnetite, as well as specular hematite, can also be a secondary magnetic mineral that typically grows in rocks long after deposition. We will cover this when we examine the remagnetization of North American cratonic rocks in the Late

Paleozoic by orogenic fluids and discuss clay diagenesis remagnetization.

The likelihood of an early reduction diagenetic event depends largely on the total organic carbon content of the sediment that, in turn, depends on the environment of the sediment's deposition. Paleomagnetists typically collect samples from fine-grained sediments because this ensures a quiet depositional environment and good alignment of the primary depositional magnetic minerals for a strong DRM or pDRM. The sediments and sedimentary rocks targeted by paleomagnetists include those deposited in deep and near-shore marine, lake and river settings. The marine and lake sediments usually have magnetite as the primary magnetic mineral. The terrestrial river sediments can have magnetite as the primary magnetic mineral, but many ancient fluvial sediments have become red beds with both primary and secondary hematite as the dominant magnetic minerals.

IRON-SULFATE REDUCTION DIAGENESIS

In marine and lake sediments with organic material, iron-sulfate reduction diagenesis or pyritization can occur relatively soon after deposition. This process can then create an early CRM through the growth of secondary magnetic minerals that formed soon after deposition, but not always. In marine sediments the total organic carbon content associated with reduction diagenesis lies within the range 1–2% (Japan Sea, northern California coastal margin, Oman margin). Lake sediments typically have higher total organic contents of 1–10%; in some cases organic contents can be as large as 20% (Bohacs *et al.* 2000). Sulfate reduction diagenesis is biologically mediated (Roberts & Weaver 2005). The *Desulfovibrio* microbe, a Gram negative sulfate-reducing bacterium, can be important in this reaction (Tarling 1999).

The chemical reactions that occur as a result of bacterial decay of organic matter in marine and lake sediments cause the dissolution of magnetite particles, followed by the creation of ferromagnetic Fe sulfides and finally the end product: non-ferromagnetic pyrite (FeS_2). The diagenetic reactions include much more than just iron and sulfate reduction, but those are the reactions important to the diagenesis of magnetic minerals. A more complete context for iron and sulfate reduction would need to include the reactions (in

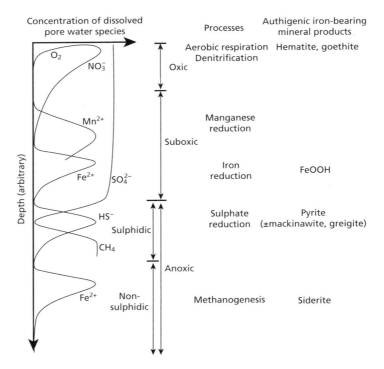

Fig. 6.1 Reduction diagenesis in recent marine sediments from Roberts & Weaver (2005). Reprinted from *Earth and Planetary Science Letters*, 231, AP Roberts and R Weaver, Multiple mechanisms of remagnetization involving sedimentary greigite (Fe3S4), 263–277, copyright 2005, with permission from Elsevier.

order) of aerobic oxidation, followed by nitrate reduction, manganese reduction, Fe oxide reduction (in which magnetite is dissolved) and finally sulfate reduction (Fig. 6.1; Froelich *et al.* 1979; Berner 1981; Kasten *et al.* 2003).

The importance of Fe oxide reduction to the dissolution of magnetite in marine sediments was first recognized in pioneering studies by Karlin & Levi (1983) and Karlin (1990), who observed a decrease in the magnetic intensity in hemipelagic marine sediments from the Gulf of California and from coastal Oregon in the top meter of the sediment column. Iron oxide reduction puts Fe^{2+} into solution that reacts with the sulfide already in solution due to sulfate diagenesis, to form a sequence of iron sulfides with the final product being pyrite. Pyrite does not carry a remanence (it is paramagnetic) so it is not important in paleomagnetic studies; however, one of its precursor minerals (greigite) is ferrimagnetic and the Fe sulfide analogue of magnetite. Its formula is Fe_3S_4, similar to magnetite

(Fe_3O_4), with sulfur taking the place of oxygen in the crystal lattice. It is an inverse spinel and cubo-octohedral like magnetite and strongly magnetic like magnetite, having intrinsic magnetizations that are about 25% of that of magnetites. Because of its strong magnetization, greigite has coercivities similar to magnetite. This makes greigite difficult to distinguish from magnetite in alternating field demagnetization studies of marine and lake sediments. The only distinctive way of distinguishing greigite from magnetite is by heating. Greigite, like other magnetic Fe sulfides (e.g. pyrrhotite), decreases in intensity at temperatures near to 300–350°C while magnetite will remain strongly magnetic up to its Curie temperature of 580°C.

When paleomagnetists notice a decrease in magnetization occurring at c. 300–350°C during thermal demagnetization, their first thought is that magnetic iron sulfides are important carriers of magnetic remanence in a sedimentary rock. This observation immediately raises the specter of a secondary diagenetic remanence. We will however see that studies in the literature are evenly divided about whether the secondary diagenetic remanence occurs so soon after deposition that it can be considered 'syn-depositional' or much later after deposition. Certainly, the organic matter diagenesis in marine and lake sediments is observed to occur near to the top of the sediment column (in about the top meter) so the reactions that create greigite should occur soon after deposition. In various papers (Roberts & Weaver 2005; Rowan *et al.* 2009), Roberts provides evidence of much later stage greigite magnetizations including SEM observations of greigite growing on pyrite (Roberts *et al.* 2005; Sagnotti *et al.* 2005; Jiang *et al.* 2001a), suggesting late-stage diagenesis forming greigite. Mackinawite ($Fe_{1+x}S$) is also one of the precursor minerals formed in the reactions that eventually lead to pyrite. It is not ferromagnetic and is therefore not of interest to paleomagnetists.

One question that should be addressed is: why doesn't all organic matter diagenesis lead to the final product pyrite, which is not ferromagnetic? Why is greigite observed in so many anoxic sedimentary rocks if it is only a precursor mineral on the way to pyrite? Roberts & Weaver (2005) point out that the conditions for the preservation of greigite in the sedimentary record are abundant ferrous iron in solution, but limited availability of sulfide in solution. For this reason, lake waters that have limited amounts of sulfate tend to have greigite preserved during lake sediment diagenesis. Another reason that greigite is more likely to be preserved in the sedimentary record is a relatively fast sediment accumulation rate that will push sediments through the diagenetic zone before the reactions can run to completion and form pyrite from the dissolved magnetite. For this reason greigite is often observed in hemipelagic near-shore marine depositional environments. Plentiful greigite has also been observed in paleomagnetic studies of both marine and lake sediments (Table 6.1) so its preservation is not unusual. When greigite was first observed as an important ferromagnetic mineral several decades ago, it was thought that it would not be a significant remanence carrier because it was metastable and would oxidize easily after sampling, during transport and storage of the paleomagnetic samples. However, it has now been observed in many paleomagnetic studies (see Table 6.1) and has even been observed in Cretaceous marine sedimentary rocks (Reynolds *et al.* 1994).

Paleomagnetists also recognize that pyrrhotite is a potentially important iron sulfide remanence carrier in sedimentary rocks. Like greigite, pyrrhotite shows a decrease in magnetic intensity at temperatures between 300 and 350°C during thermal demagnetization. It comes in monoclinic and hexagonal forms; monoclinic pyrrhotite (Fe_7S_8) is ferrimagnetic, due to missing Fe cations in its crystal lattice, and strongly magnetic like greigite and magnetite. Hexagonal pyrrhotite (Fe_9S_{10}) is antiferromagnetic and is magnetic, like the monoclinic form, if there are missing cations in the crystal lattice. Pyrrhotite is usually found in igneous and metamorphic rocks. It can form during sulfate diagenesis in sediments, but it forms exceedingly slowly at temperatures less than 180°C, making it less important in sediment diagenesis (Horng & Roberts 2006). If it is identified in sediments, it is more likely to be a depositional magnetic mineral (Horng & Roberts 2006).

The critical question for paleomagnetists is how soon after deposition and over what time interval is the CRM, carried by the secondary iron sulfide minerals, created by reduction diagenesis. Table 6.1 lists 29 studies from the literature that have identified greigite or secondary iron sulfides as an important remanence carrier. For 10 localities, the CRM carried by greigite or iron sulfides is observed to be 'syn-depositional' in age. This observation is consistent with the observation that sulfate and iron reduction diagenesis usually occur in about the top meter of the sediment column.

For 10 localities evidence is however given of late-stage diagenesis of the CRM-carrying greigite. In some cases there is direct observation of greigite growing on secondary pyrite grains, clear indication of post-early diagenesis greigite formation. Two studies show evidence for both early and late-stage greigite formation (Table 6.1). Almost all these studies are of marine sediments; just two studies concern lake sediments and one is of terrestrial gravels and clays. In the Rowan *et al.* (2009) study of Pleistocene–Holocene marine sediments from northern California and Oman, some actual time constraints are put on greigite formation. The sediments are hemipelagic with moderate sediment accumulation rates of 7 cm/kyr. The age of the greigite CRM is offset from the depositional age by tens of thousands of years and the record is smoothed over by tens–hundreds of thousand years.

This sad picture should however be contrasted with the study of Kodama *et al.* (2010) of the Eocene marine marls of the Arguis Formation in Spain which observed precessional-scale (20 kyr) Milankovitch cycles in the concentration of magnetic minerals that included both

Table 6.1 Secondary Fe sulfides: late versus early diagenesis

Locality	Lithology	Mag Min	Age	Diagenesis	Reference
Oman/North California	Marine	Greigite	Pleistocene–Holocene	Late	Rowan et al. (2009)
New Zealand/Crostolo R. section	Marine	Greigite	Neogene	Late	Rowan & Roberts (2006)
Italy	Marine	Greigite	Olduvai trans.	Late	Roberts et al. (2005)
Taiwan	Marine	Greigite	Pliocene	Late	Roberts & Weaver (2005)
Antarctica/Ross Sea core-CRP-1	Glacio/ marine	Greigite	Miocene	Late	Roberts & Weaver (2005)
Paleo Tiber valley, Italy	Gravels/clay	Greigite	Mid-Pleistocene	Late	Roberts & Weaver (2005)
Antarctica/Ross Sea core-CRP-1	Glacio/ marine	Greigite	Mid-Pleistocene	Late	Sagnotti et al. (2005)
Sakhalin, Russia	Marine	Greigite/Pyrrhotite	Miocene	Late	Weaver et al. (2002)
Okhotsk Sea cores	Marine	Greigite	Holocene	Early	Kawamura et al. (2007)
Montalbano, Janico section Italy	Marine	Greigite	B-M transition age	Late	Sagnotti et al. (2010)
Northern Italy	Marine	Greigite	15 Ma	Early	Huesing et al. (2009)
Tortonia ref sec, north Italy	Marine	Greigite	Miocene	Early	Huesing et al. (2009)
Carpathian foredeep	Marine	Greigite	Mio-Pliocene	Early	Vasiliev et al. (2007)
Dead Sea core	Saline Lake	Greigite	Recent	Early	Franke et al. (2007)
Crag Basin, East Anglia, UK.	Marine/intertidal	Greigite	Pleistocene	Early	Maher & Hallam (2005)
Coastal NJ	Marine	Greigite/Magnetite	B-M transition age	Both Early and Late	Oda & Torii (2004)
Gutingkeng Formation, Taiwan	Marine mudstones	Greigite	Early Pliocene	Late	Jiang et al. (2001b)
Norfolk coast, UK	Estuarine	Greigite	Early Pleistocene	Early	Hallam & Maher (1994)
North California	Continent. borderland marine	Iron sulfides	Recent	Early	Leslie et al. (1990)
Russia, Lake Aslikul	Lacustrine	Iron sulfides	Recent	Early	Nurgaliev et al. (1996)
Pyrenees section, Spain	Marine marls	Iron sulfides	Eocene	Early	Kodama et al. (2010)
Crostolo section, north Italy	Marine	Iron sulfides	Olduvai trans.	Early	Tric et al. (1991)
Awatere Group, South Island, New Zealand	Siliciclastic	Greigite/pyrrhotite	Late Neogene	Early	Roberts & Turner (1993)
Gutingkeng Formation, Taiwan	Mudstone	Magnetite/pyrrhotite/greigite	Pleistocene	Greigite is Late	Horng et al. (1998)
North Sea cores	Marine	Iron sulfides	Cenozoic	Late	Thompson & Cameron (1995)
E. Mediterranean Sea	Marine	Iron sulfides	Plio-Pleistocene	Late	Richter et al. (1998)
Korea Straight	Marine	Iron sulfides	Middle Holocene	Both Early and Late	Liu et al. (2004)
Dolomites, Italy	Marine	Greigite	Triassic	Early	Spahn et al. (2011)

depositional magnetite and secondary iron sulfides. Larrasoana *et al.*'s (2003) study of the same rocks demonstrated similar polarities for the Fe sulfides and magnetite in a magnetostratigraphic study. Huesing *et al.*'s (2009) high-resolution cyclostratigraphic study of marine sediments from northern Italy showed good correlation between a magnetostratigraphy recorded by greigite with the biostratigraphy and argues strongly for an early 'syn-depositional' age of the greigite. Spahn *et al.*'s (2011) magnetostratigraphy from the Triassic limestones of the Dolomites in northern Italy showed similar directions carried by greigite and magnetite, suggesting an early nearly depositional age for the secondary greigite.

The presence of secondary iron sulfide magnetic minerals in a sedimentary rock does not necessarily spell doom for nearly syn-depositional age of the remanence, although this possibility should be rigorously checked with whatever means are possible such as reversal and fold tests.

EARLY DIAGENESIS IN TERRESTRIAL RED BED SEDIMENTARY ROCKS

We have already discussed the magnetization (Chapter 2) and magnetic anisotropy of red sedimentary rocks (red beds; Chapters 4 and 5). We have seen that red beds are very important sources of paleomagnetic data, particularly for the Mesozoic, Paleozoic and earlier in Earth history. We also mentioned that in the 1970s and 1980s there was a good deal of controversy in paleomagnetic circles about whether the characteristic magnetization of sedimentary red beds was due to a primary DRM or a secondary CRM. Butler (1992) does an excellent job of summarizing the 'red bed controversy'; many of the main points that he made in the early 1990s are still relevant today.

Before repeating those points here to aid our discussion, it is important to realize that the magnetization of red beds is due to hematite (Fe_2O_3) and that typically for red beds there are two forms of hematite (and hence two magnetizations) in red beds. One is the obvious pigmentary hematite that gives the red bed its characteristic color. Many studies, particularly by T.R. Walker and his colleagues (e.g. Walker 1967, 1974; Walker *et al.* 1978, 1981; Larson *et al.* 1982), indicate that the pigmentary hematite is submicron in size and most likely due to the post-depositional inter-stratal alteration of primary Fe silicate minerals. Thermal demag-

netization of red beds however suggests that a larger-grained hematite carries the highest temperature magnetization in the rocks, and it may be different in direction from the magnetization carried by the pigmentary hematite. This magnetization is typically attributed to specular hematite (or specularite). The age of this hematite is what is at issue in the 'red bed controversy', i.e. whether it is depositional or secondary in age.

Butler (1992) points out that there is evidence that in some red beds the specular hematite carries a DRM. These rocks are likely to be more 'mature' sediments and relatively coarse-grained clastic rocks. Butler characterizes this interpretation as the 'minority view' among paleomagnetists. The 'majority view', according to Butler's reading of the literature in the early 1990s, was that red beds carry a CRM that was created by early diagenesis of the sedimentary rocks. The early diagenetic CRM of red beds, carried by specular hematite, can still be a paleomagnetically useful signal if it is formed within 10^3–10^4 years of deposition, according to this view. For a Mesozoic or Paleozoic rock that is tens or hundreds of millions of years old, this is an inconsequential difference from the true depositional age.

One of the new pieces of paleomagnetic information about red bed magnetization that has come to light since Butler's (1992) book is that the magnetic fabric, particularly the anisotropy of the hematite carrying the high-temperature characteristic remanence, has maximum and intermediate principal axes lying in the bedding plane and the minimum principal axis perpendicular to the bedding plane. This is a fabric characteristic of depositional or compaction processes, or the result of both, acting on the magnetic minerals in a rock. The other piece of new information is that both the anisotropy-based inclination-correction and EI correction techniques have identified shallow inclinations in red bed units. This new information is relevant to our understanding of the acquisition of red bed remanence.

Before we integrate our new information about red bed remanence, it is useful to briefly summarize the understanding of red beds that has evolved in the literature. The early prevailing view in the 1960s and 1970s was that red beds formed in desert conditions. Meticulous work by Walker (1967) on recent alluvial deposits in Baja California and Walker *et al.* (1978, 1981) and Larson *et al.* (1982) on more ancient red beds suggested that inter-stratal alteration of Fe silicate minerals was the cause of the red coloration of red beds. Van Houten's (1973) review article and Turner's (1980) book, *Continental Red Beds*, are good proponents of the view that red beds formed in arid environments and were the result of diagenesis – both early diagenesis in the post-depositional history of rocks and late diagenesis that could persist for tens of millions of years after deposition. Obviously this view suggested that the CRM of red beds would be secondary, and probably not of much use paleomagnetically. Walker (1974) studied recent deposits in Orinoco which he suggested would turn into classic red beds given enough time, and indicated that red beds could form from clastics deposited in humid environments and were not limited to an arid environment of deposition.

A new idea about red beds was introduced by Dubiel & Smoot (1994) soon after the publication of Butler's (1992) book. Red bed formation was favored by a monsoonal climate; alternating wet and dry seasons and savannah-type vegetation would cause the diagenesis needed to create red beds observed in the geologic record. Seasonal aridity would cause the growth of hematite geologically soon after deposition, probably when the water table was deep in the dry season. Parrish's (1998) book echoes the role of monsoonal seasonality in red bed formation. In their study of the Late Cretaceous Pozo Formation red beds of the Salinian block in California, Whidden *et al.* (1998) argue that the Pozo magnetization is due to early diagensis, i.e. pedogenesis within 10^5 years of deposition, in a seasonally arid climate. The observation of a magnetostratigraphy in the Pozo red beds supports an early age of their magnetization. Whidden *et al.* argue that the early formation of the hematite carrying the Pozo magnetization occurred above the water table in an overbank deposit. They provide good petrographic and paleomagnetic evidence in support of their interpretation.

The first evidence that the remanence-carrying hematite in red beds had a characteristic compaction and/or depositional fabric came from the Mississippian Mauch Chunk Formation and Cretaceous Kapusaliang Formation red beds (Tan & Kodama 2002; Tan *et al.* 2003). Either stepwise dissolution of the magnetic minerals in chemical demagnetization coupled with AMS measurements or high field laboratory-applied magnetizations (IRMs) parallel and perpendicular to the bedding plane showed that the specular hematite grains that carried the high-temperature remanence

revealed a bedding-parallel oblate fabric. Earlier AMS measurements of red beds (Bossart *et al.* 1990; Garces *et al.* 1996) had shown similar fabrics, but the Mauch Chunk and Kapusaliang studies showed that the hematite carrying the ancient remanence had depositional/compaction fabrics. This work had been performed as part of an anisotropy correction for inclination shallowing and did reveal shallowing for these red bed units. Subsequently, as indicated in Chapter 5, the anisotropy-based and EI techniques showed shallow inclinations for a good number of red bed units from all over the world. Both the anisotropy data and the shallow inclinations support an early magnetization age for red beds, if not a DRM then an early CRM, that could be affected by compaction.

Sheldon (2005) has presented an interesting new twist in the red bed formation story. Sheldon points out that red beds had traditionally been interpreted paleoclimatically to indicate desert conditions (Walker 1967), but more recently had been interpreted to indicate monsoonal, seasonal aridity (Dubiel & Smoot 1994; Parrish 1998). Both of these interpretations have informed paleomagnetists as they have attempted to understand the timing and manner by which red beds are magnetized. Sheldon (2005) studied the Permian red beds of Cala Viola, Sardinia and showed through elemental analysis of chemical weathering, root traces and trace fossils, as well as pedogenic carbonate and salts, that the red beds formed pedogenically due to good drainage and not because of any particular paleoclimate conditions. Their formation could not be attributed to aridity or even seasonal aridity during a strong monsoon. Sheldon concludes that the red color of the Cala Viola pedosols is primarily due to the hydrological conditions in which they formed. To a paleomagnetist this would suggest that the hematite carrying the Cala Viola red bed remanence was formed by early diagenesis, not by extended late diagenesis, and is consistent with the picture that has developed from remanence and anisotropy measurements.

To develop a model for diagenesis relevant to the magnetization acquisition of red beds, it's best to summarize all the observations we have about red beds. First, red beds are typically continental sediments deposited in fluvial, lacustrine or eolian conditions. Some red beds are formed from paralic or deltaic nearshore marine sediments, but it is more likely that they were deposited terrestrially. Metcalfe *et al.* (1994) suggest that they are typically formed in rapidly subsiding fault-bounded basins. The molasse sediment of the classic Paleozoic red beds of the Appalachians would fit this description, as would the Mesozoic basins of the North American east coast (e.g. Newark and Hartford basins).

Petrographic examination of classic, paleomagnetically important red beds indicate that inter-stratal alteration of ferromagnesian silicates and oxides (pyroxenes, biotites, amphiboles, olivine, magnetite, etc.) was the primary source of the pigmentary hematite in the rocks. Iron oxyhydroxides, which form during diagenesis, are the source of the hematite through dehydration reactions. In an early study of red bed genesis, Walker (1967) reported that the distinctive red bed coloration and inter-stratal alteration of alluvium in northeastern Baja California was not evident until the sediments were Pliocene in age (2–5 Ma). Pleistocene sediments at the ground surface were red due to soil formation but were more typically yellow/red colored. This observation is consistent with a survey of the 2005 version of IAGA Global Paleomagnetic Database (McElhinny & Lock 1990) that shows the youngest reported red sediments in the database are 2–3 million years old (the red clays of the Dnieper River sediments, Ukraine and the Kuban River sediments, Russia). The youngest rocks designated as red beds are the 3–5 Ma sediments of the Karanak Group from Tajikistan.

Thermal demagnetization of red bed remanence typically shows a distributed unblocking-temperature spectrum at intermediate temperatures (300–650°C), usually attributed to the fine-grained pigmentary hematite, and a narrow unblocking-temperature spectrum at the highest temperatures (660–680°C) near to the highest unblocking temperatures for hematite (Neel temperature *c.* 685°C). The narrow highest unblocking temperatures are usually attributed to the specular hematite that carries the characteristic remanence of a red bed (Fig. 6.2). Red beds can carry a well-defined magnetostratigraphy with polarity zones confined to strata. Most recently, the characteristic remanence-carrying hematite grains in red beds are observed to have a compaction and/or depositional fabric (Fig. 5.9) and to have suffered from inclination shallowing, similar in magnitude to the shallowing observed in red bed laboratory re-deposition experiments (Tauxe & Kent 1984; Tan *et al.* 2002).

These observations suggest that, for paleomagnetically important red beds, the remanence is either depositional or very early diagenetic (within 10^5 years if not

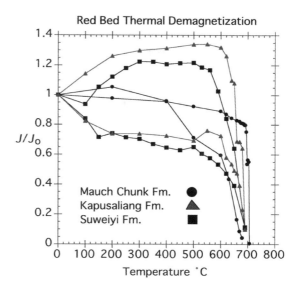

Fig. 6.2 Intensity decrease during thermal demagnetization of NRM for representative red bed magnetizations. Data from Cretaceous Kapusaliang, the Mississippian Mauch Chunk and the Paleocene Suweiyi formations.

10^3–10^4 years of deposition). If it is diagenetic, the hematite carrying the remanence grew early enough after deposition to acquire a burial compaction fabric. Another possibility is that the hematite formed diagenetically inherited the compaction/depositional fabric of the hematite's precursor minerals, although this scenario would not be sufficient to explain the bedding-confined polarity zones that allow good magnetostratigraphies in red beds.

The question raised in Butler's book is still unresolved at the end of the analysis in this section: what is the source of the specular hematite in red beds that carries their characteristic high-unblocking-temperature remanence? Are the specular hematite grains depositional or secondary? I would argue that the evidence should swing paleomagnetists more toward the 'minority view' in Butler's book, that is, that the specular hematite is likely to be depositional in many cases, but that there can certainly be cases where specularite is formed from diagenesis (even diagenesis that has occurred over hundreds of thousands or even a million years since deposition). The age of red bed remanence needs to be addressed carefully on a case-by-case basis and informed by anisotropy measurements and tests for inclination shallowing.

LATE DIAGENESIS AND REMAGNETIZATION

There are several processes that cause a late-stage diagenesis, tens to hundreds of millions of years after deposition of a sedimentary rock, that through the growth of secondary magnetic minerals create a 'classic' remagnetization in the rocks. The remagnetization may be partial or complete, either totally resetting the rock's magnetization or only adding another component of magnetization that can hide the primary magnetization of the rock. The three major processes outlined in the literature over the last several decades are: basinal fluid flow associated with orogenesis, clay diagenesis and hydrocarbon maturation. The process that by far has received the most attention is basinal fluid flow associated with diagenesis, so it will be discussed first. It is anticipated that readers can also refer to Van Der Voo & Torsvik's review on the subject which, at the time of writing (March 2012), is due to be published.

Basinal fluid flow during orogenesis

The Late Paleozoic remagnetization event, primarily in North America, is one of the most extensively studied and documented remagnetizations in the literature. It is the paleomagnetic consequence of the 'squeegee tectonics' proposed by Oliver (1986) in which fluids were squeezed through rocks during orogenesis. The remagnetization occurred during the Kiaman Permian-age reversed-polarity superchron so remagnetized rocks typically have a very shallow, nearly due south, paleomagnetic direction in North America. The geographic extent of the remagnetization is typically in Appalachian Paleozoic rocks from eastern North America and in the flat-lying Paleozoic rocks of the mid-continent. McCabe & Elmore (1989) provide an excellent review paper summarizing results up to the late 1980s. The major points that McCabe and Elmore provide about the Kiaman age remagnetization in North America are as follows.

1. Typically carbonate rocks are totally remagnetized and the magnetization is carried by diagenetic magnetite. In the Appalachians the magnetization is post- or syn-folding in age in the carbonates (Fig. 6.3). The magnetite carrying the remagnetization appears to be framboidal in morphology, suggesting that it is diagenetically grown (Fig. 6.4).

2. Red siltstones and sandstones carry a partial remagnetization. The lower-unblocking-temperature magnetization, perhaps associated with the pigmentary, submicron hematite in the rocks, carries a syn- or post-folding Kiaman-age overprint on a high-unblocking-temperature primary characteristic magnetization carried by specular hematite.

3. Despite some controversy in the 1980s about the possibility that the remagnetization was due to a thermoviscous effect, i.e. elevated temperatures over long periods of time, the evidence of diagenetic morphologies for the magnetite particles in the carbonate rocks, along with their syn- or post-folding magnetization ages in the Appalachians, argued for basinal fluid flow as the cause of a CRM magnetization. From some fluid inclusion studies, the fluid was observed to be warm brine which was rich in potassium. It would have caused clay diagenesis that in turn provided the Fe for the secondary magnetite. Interestingly enough, the same fluids would have had to produce secondary pigmentary hematite in the siltstones and sandstones at the same time, so the chemical changes were lithologically dependent.

4. There was an initially puzzling observation of magnetic fabric data from remagnetized carbonates that appeared to be primary. The magnetic fabric for the remagnetized carbonates, the prime example being the Onondaga Limestone, was measured by magnetic susceptibility (AMS). The evidence for a primary magnetic fabric for obviously remagnetized rocks is important to this book, since in the previous section we argued that primary magnetic fabrics suggested DRMs or very early CRMs for red beds. We will return to the primary-seeming AMS fabrics in the Onondaga Limestone later in this section.

Since the publication of the McCabe & Elmore (1989) review article, there has been additional work done on the Late Paleozoic remagnetization event. Jackson (1990) and Jackson *et al.* (1992) conducted detailed rock magnetic studies of remagnetized carbonates and made some important observations. The most important was that the remagnetized carbonates had a characteristic signature in their hysteresis plots where the magnetization of a rock sample is cycled slowly back and forth in strong magnetic fields that saturate the

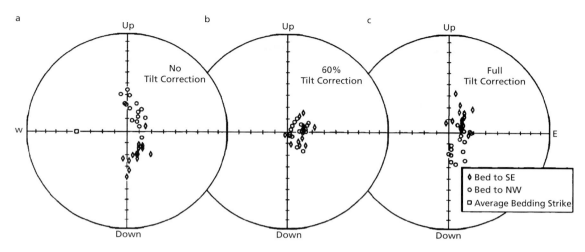

Fig. 6.3 Stepwise fold test for the remagnetized Helderberg Formation carbonate rocks. The site means cluster best when the fold limbs are 'unfolded' by 60% of their bedding tilt, indicating that the rocks were remagnetized during folding. C McCabe and RD Elmore, The occurrence and origin of Late Paleozoic remagnetization in the sedimentary rocks of North America, *Review of Geophysics*, 27, 4, 471–494, 1989. Copyright 1989 American Geophysical Union. Reproduced by permission from American Geophysical Union.

Fig. 6.4 Magnetite framboids assumed to be diagenetic separated from the remagnetized Bonneterre Dolomite (McCabe *et al.* 1987). Scale bar is 10 μm. C McCabe, R Sassen and B Saffer, Occurrence of secondary magnetite within biodegraded oil, *Geology*, 15, 7–10, 1987. Copyright 1987 American Geophysical Union. Reproduced by permission from American Geophysical Union.

magnetization. The hysteresis plots were described as 'wasp-waisted', i.e. their middle sections were narrower than their upper and lower sections (Fig. 6.5).

Jackson interpreted the wasp-waisted plots as due to a mixture of large (multi-domain) and small (single domain) magnetic grains (magnetite in the case of the remagnetized carbonates). The wasp-waisted hysteresis signature was subsequently used by workers to help identify remagnetized carbonate rocks. Jackson also showed that a high ARM/SIRM ratio (>0.10) could also be used as evidence for remagnetization.

Lu & McCabe (1993) and Sun *et al.* (1993) both attacked the problem of primary appearing AMS fabrics in a remagnetized carbonate, the Onondaga Limestone, first reported by Kent (1979). Lu and McCabe measured remanence anisotropies, both AAR and AIR, for remagnetized carbonates from the southern Appalachians and found primary appearing fabrics for the AAR in low-coercivity (large) magnetic grains and secondary tectonic fabrics for the AIR in high-coercivity (small) magnetic grains.

Sun *et al.* (1993) approached the problem slightly differently, examining the AAR of many different

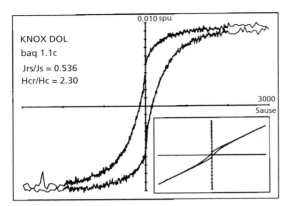

Fig. 6.5 Hysteresis loop of remagnetized Knox Dolomite from Jackson (1990) showing wasp-waisted behavior of a remagnetized carbonate rock. M Jackson, Diagenetic sources of stable remanence in remagnetized Paleozoic cratonic carbonates: A rock magnetic study, *Journal of Geophysical Research*, 95, B3, 2753–2761, 1990. Copyright 1990 American Geophysical Union. Reproduced by permission from American Geophysical Union.

remagnetized North American carbonates. They found that strongly remagnetized carbonates had nearly horizontal minimum principal axes, indicating a secondary (perhaps tectonic) fabric and weakly and non-remagnetized carbonates that typically had primary fabrics with minimum axes perpendicular to the bedding plane. The explanation for the initially observed primary fabrics in the Onondaga now becomes clearer; AMS preferentially measures the fabric of the coarser magnetic particles, perhaps the depositional magnetite in the rocks. AAR shows that these magnetic particles appear to have a primary fabric. The secondary fine-grained single-domain magnetite particles have a secondary magnetic fabric that the AMS measurements may have missed. Remanence anisotropy is therefore superior to AMS for detecting the fabrics of subpopulations of the magnetic grains in a rock. The remanence anisotropy of the high-unblocking-temperature hematite in red beds can be a good indicator of their primary origin.

One of the problems that McCabe and Elmore pointed to was the lack of any petrographic evidence about the partial remagnetizations observed in the siliciclastic rocks suffering from Late Paleozoic remagnetization. Lu *et al.* (1994) studied the hematite remagnetization in the Silurian Rose Hill Formation

red beds and showed, petrographically, that the hematite carrying the secondary magnetization is authigenic and derived from alteration of ferromagnesian silicates.

Cox *et al.* (2005) conducted a detailed study of the remagnetized Devonian red beds of the Hampshire Formation in West Virginia. They studied this unit where it outcropped in the more interior Alleghanian Plateau and in the folded Valley and Ridge province. Petrographic examination showed authigenic specular hematite cement and submicron hematite pigment. Fluid inclusion studies did not show warm orogenic fluids, as had been shown for other units in the remagnetized Appalachians, but instead suggested that the secondary hematite was formed by a mixture of methane-saturated formational fluids and meteoric fluids. Cox *et al.* see a two-step process for the remagnetization of the Hampshire. First, methane-fluids, perhaps driven through the rocks by orogenesis, caused dissolution of depositional Fe-oxide magnetic minerals putting Fe into solution. Meteoric fluids that arrived during subsequent post-orogenic uplift of the rocks then caused oxidation of the Fe in solution to precipitate authigenic hematite. Although the authors state that the standard warm orogenic fluid model cannot be invoked to explain the secondary magnetization of these red beds, clearly late-stage fluid flow (in this case involving hydrocarbons) could have caused the remagnetization.

Stamatakos *et al.* (1996) give important evidence for the timing of the Kiaman remagnetization and its rapid occurrence. They examined the age of remagnetization from fold tests applied to remagnetized rocks along a transect that crossed the folded rocks of Pennsylvania. They found that rocks in the hinterland (eastern PA) had post-folding magnetizations while rocks closer to the foreland (central and western PA) had syn- and even pre-folding Kiaman remagnetizations (Fig. 6.6). This observation suggested that orogenic fluids moved quickly through the rocks over 20 million years during the formation of the Appalachian fold and thrust belt, catching a 'snapshot' of the folding at different stages in development. This observation was indirect evidence of a fluid cause of the remagnetization, even in siliciclastic rocks since Stamatakos *et al.* used both carbonates (Greenbriar, Helderberg, Clinton, Trenton, and Allentown/Leithsville) and siliciclastic red beds (Mauch Chunk, Catskill, Andreas, Bloomsburg, Rose Hill and Juniata) in their study.

Additional evidence for a basinal fluid cause of the remagnetization comes from McCabe *et al.*'s (1989)

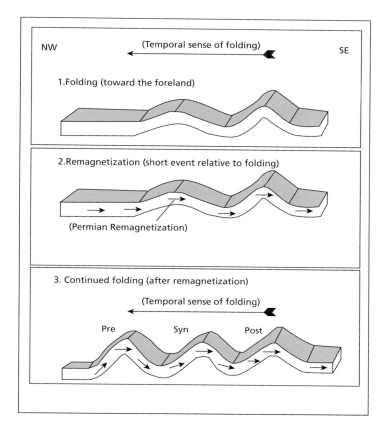

Fig. 6.6 Model for the formation of pre-, syn- and post-folding remagnetizations in the Appalachians due to the Kiaman age remagnetization event (Stamatakos *et al.* 1996). Based on the paleomagnetic poles from the remagnetized rocks, the remagnetization occurred over a relatively short 20 million year period between 275 and 255 Ma.

study of transects of remagnetized rocks in New York and in Tennessee/Alabama. They observed that the remagnetization was related to the degree of diagenetic alteration of clay minerals and that Fe was derived for secondary magnetic minerals from the illitization of smectites. We will consider this mechanism in more detail in the following section on remagnetization caused by clay diagenesis. In addition, McCabe *et al.* (1989) saw evidence of what they called a remagnetization 'shadow' where the postulated basinal fluids went deep, avoiding shallower, younger Paleozoic rocks, because the highly impermeable Chattanooga Shale deflected the fluids. Elmore *et al.* (1993) examined the spatial relationships of remagnetization in the Ordovician Viola Limestone of Oklahoma to veins in the rock. They found pervasive remagnetization in sec-

ondary magnetite of the Viola, but that a remagnetization residing in secondary hematite decreased with distance from the veins. Fluid inclusion and ^{87}Sr isotopic studies showed that the secondary fluids purportedly causing the remagnetization were warm, saline and radiogenic.

Finally, studies since McCabe & Elmore's (1989) review article show that there are examples of: Late Paleozoic Kiaman-age remagnetized rocks in miogeosynclinal carbonate rocks from Nevada, Utah and California carried by secondary magnetite (Gillett & Karlin 2004); remagnetized Precambrian rocks in the Rocky Mountains with the magnetization carried by secondary hematite (Geissman & Harlan 2002); and remagnetized Lower Carboniferous carbonates from the Craven Basin of northern England (McCabe &

Channell 1994) that were deformed during the Hercynian orogeny (contemporaneous to the Alleghanian in North America). All this evidence shows how widespread the Kiaman age remagnetization was; despite being concentrated in eastern and mid-continent North America, its effects could be seen as far away as the North American west and Hercynian England.

Fairly recent paleomagnetic studies are still affected by this large-scale remagnetization. Cederquist *et al.* (2006) and Hnat *et al.* (2009) showed that carbonate and siliciclastic rocks from the Tennessee and Pennsylvania salients are either totally or partially remagnetized in the Late Paleozoic and the remagnetized directions show no evidence of oroclinal bending, so that any bending that occurred of a primarily straight mountain belt must have occurred before the Late Paleozoic.

Burial diagenesis and illitization

The primary mechanism proposed for remagnetization due to the very general term 'burial diagenesis' is clay diagenesis. More specifically, this entails the change of detrital smectite clays to secondary illite clays during burial, called illitization. Illitization provides the iron necessary for the formation of secondary magnetite that, in turn, would acquire a CRM as the newly formed magnetite particles grew large enough to block-in a stable magnetization. There are only a handful of studies that support clay diagenesis as a cause of remagnetization. Most are based on a 'presence–absence' argument in which the magnetizations of equivalent-age rocks are compared at two localities where the rocks have experienced illitization at one and detrital smectite is still present in the rocks at the other. If the magnetizations are different at the two localities, the argument is made that the magnetization of the illitized rocks is secondary.

There are only three study areas that have been used to support clay diagenesis remagnetization: Scotland, Montana and the Vocontian trough in France. The study of Jurassic sedimentary rocks in Skye, Scotland is the best example of the presence–absence argument (Woods *et al.* 2002). The rocks with detrital smectite and no evidence of illitization have a weak magnetization, while equivalent-age rocks at a different location that have clearly undergone illitization have stronger multi-component magnetizations. The

highest-temperature magnetization component is carried by magnetite and interpreted to be a CRM.

Gill *et al.* (2002) compared the magnetization of equivalent-age rocks in the thrusted, disturbed belt of Montana with relatively undeformed rocks in the Sweetgrass Arch to the east. The rocks in the Sweetgrass Arch have unaltered detrital smectite clay, while the Montana deformed rocks have been illitized. The deformed rocks have a stronger, more stable magnetization carried by magnetite that is interpreted to be a CRM.

In the Vocontian trough of southeastern France, Jurassic and Cretaceous carbonates that have experienced illitization have a secondary pre-folding normal polarity magnetization carried by magnetite (Katz *et al.* 1998, 2000). The magnetization is interpreted to be a CRM since the burial depths were not great enough to cause a thermoviscous resetting of a primary magnetization. Where the detrital magnetite is still present in significant amounts, the CRM magnetization is either absent or very weakly developed. In the Vocontian rocks, the intensity of the clay diagenesis CRM increases as the degree of illitization increases. Sr, C and O isotopic analyses show that an additional secondary reversed-polarity component of magnetization in rocks closer to the Alps may be due to orogenic fluids, but that rocks further from the Alps have only been in contact with ancient seawater. This suggests that, for these rocks, illitization is the cause of the magnetite CRM.

In contrast to this is Zwing *et al.*'s (2009) study of rocks from the Rhenish Massif in Germany. In a detailed study, Zwing *et al.* used K–Ar geochronologic techniques to date the illitization and observed two illitization events. Based on the age of the illitization events and the age of the secondary CRMs, Zwing *et al.* argue against illitization as the cause of the remagnetization they observe in the rocks. Although they observe that the age of remagnetization is equivalent in age to the younger illitization event in Upper Devonian rocks, it is not equivalent in age to Middle Devonian rocks. Rare Earth element (REE) studies indicate that orogenic fluids are not implicated.

Tohver *et al.* (2008) came to a different conclusion when they dated the illitization of the clay residue using $^{40}Ar/^{39}Ar$ geochronology in carbonate rocks from the Cantabrian–Asturian arc and found it to be the same age as the secondary magnetite carrying the remagnetization of these rocks. They argued that the smectite–illite transformation was the source of the Fe for the secondary magnetite. Before illitization is widely accepted as a remagnetization mechanism, more detailed studies need to be conducted like those of Zwing *et al.* and Tohver *et al.* which date the illitization event.

Other remagnetization mechanisms: Hydrocarbon migration and dolomitization

In their comprehensive review of remagnetization, McCabe & Elmore (1989) also mention hydrocarbon migration and maturation as a cause of remagnetization in addition to their main focus on orogenically-driven fluids. McCabe & Elmore (1989) indicate that Donovan *et al.* (1979) were the first to identify a relationship between hydrocarbon migration and the possible formation of magnetite. Donovan *et al.* implicate hydrocarbon migration in the formation of secondary magnetite from the reduction of hematite in an Oklahoma oilfield although, as McCabe & Elmore (1989) indicate, the work came under close scrutiny that lead Reynolds *et al.* (1985) to suggest that the magnetite was merely the result of drilling contamination.

More recent work has documented a secondary CRM with hydrocarbon migration in two carbonate units: the Mississippian Deseret Limestone in Utah and the Pennsylvania Belden Formation carbonates in Colorado. In both units, a CRM remagnetization was identified that was the same age as the modeled time of hydrocarbon migration in the rocks (Banerjee *et al.* 1997; Blumstein *et al.* 2004). Hydrocarbon migration was also implicated in the formation of secondary magnetite, pyrrhotite and specular hematite from the reduction dissolution of hematite in Montana's Chugwater Formation red beds (Kilgore & Elmore 1989). The specular hematite carried a secondary CRM.

Both dolomitization, in which Mg is added to calcite during mudrock burial dewatering, and de-dolomitization in which the Mg is removed, can cause the growth of secondary magnetic minerals. Addison *et al.* (1985) show that in the Pendleside Formation carbonates from the Craven Basin in England, un-dolomitized limestones have magnetite with a primary magnetic fabric and that dolomitzed carbonates have a CRM residing in secondary hematite. Secondary hematite is also the culprit in de-dolomitized carbonates (McCabe & Elmore 1989).

CONCLUDING THOUGHTS: ACCURACY OF A CRM REMAGNETIZATION

Since the theme of this book is accuracy of the paleomagnetic signal, it is worth looking at the laboratory experimental work on the accuracy of a CRM. There have been several important laboratory CRM studies. Stokking & Tauxe (1990a, b) performed early experiments in which a CRM was carried by secondary hematite crystals grown from nothing. In this case, the hematite carried an accurate record of the magnetic field during their growth. When Stokking and Tauxe grew hematite in a series of experiments (i.e. 'multi-generational' hematite) in the presence of sequentially perpendicular fields, they found that the resulting CRM was complex and could be either be parallel or antiparallel to the growth fields or intermediate between the fields.

Walderhaug (1992) heated basic igneous rock samples so that strongly magnetic titanomagnetites were altered to either magnetite or hematite. The resulting CRM in the secondary hematite was intermediate in direction between the pre-existing NRM and the magnetic field during growth for some samples, and parallel to the laboratory field in other cases. Madsen *et al.* (2002) followed this work with the heating of igneous rocks with NRMs at different angles to the magnetic field during heating, mimicking a CRM acquired in a fold carrying a pre-folding magnetization. In all cases, the secondary CRM was intermediate between the CRM field and the NRM direction of the sample.

Cairanne *et al.* (2004) heated a pyrite-rich claystone at low temperatures for an extended period of time and observed the oxidation of pyrite first to magnetite and then to hematite. The samples were heated in fields of successively different polarities and the CRM acquired an accurate record of the last polarity field during the heating.

What appears to be emerging from these experiments is that, for magnetic grains that grew from 'nothing' i.e. with no magnetic precursor minerals, the resulting CRM is an accurate record of the magnetic field during growth. Since pyrite does not carry a remanence, the magnetic minerals that grow from the oxidation of pyrite can carry an accurate CRM. However, for those magnetic minerals that grow from the alteration of a pre-existing magnetic mineral phase, i.e. oxidation of magnetite to hematite or multiple generations of hematite, the NRM affects the direction of the CRM acquired by the secondary magnetic minerals. The resulting CRM is intermediate between the pre-existing NRM direction and the magnetic field direction during growth, and is not an accurate representation of the geomagnetic field.

A remagnetization carried by a CRM can therefore be accurate, but it depends on the magnetic environment in which the secondary magnetic minerals grew. In the great Late Paleozoic remagnetization of eastern North America caused by orogenically driven fluid flow, both the secondary magnetite in the carbonate rocks and the secondary pigmentary hematite in the clastic rocks probably grew from 'nothing' i.e. there were no permanently magnetic precursor minerals, so the CRM acquired is an accurate record of the geomagnetic field. This is supported by paleopoles, carried by the remagnetization, that are consistent with the North American magnetic field direction for the Late Paleozoic. The accuracy of a remagnetization should be considered on a case-by-case basis with the nature of the precursor minerals an important point to inform the interpretation of the secondary CRM.

Tectonic Strain Effects on Remanence: Rotation of Remanence and Remagnetization in Orogenic Belts

Paleomagnetists naturally gravitate to folded rocks for paleomagnetic measurements. The reason is simple: paleomagnetists can constrain the age of a rock's magnetization by using Graham's (1949) fold test, a very powerful field test that shows whether a rock's paleomagnetic vector is older, younger or the same age as the folding of a rock (Fig. 7.1). Even Graham (1949), in the first exposition of the fold test in the literature, recognized that two types of deformation can affect a rock's remanence: the rigid body rotation of a fold's limbs and the internal strain that folded rocks undergo as the rocks are bent into folds. Historically, paleomagnetists have concentrated mainly on the rigid body rotations of fold limbs, undoing them mathematically to see if the magnetization on both limbs will cluster better before, during or after the rigid body rotations are removed by simple rotation around the strike of the bedding. With the first observation of 'syn-folding' magnetizations (Scotese *et al.* 1982), i.e. when magnetizations cluster best during partial untilting of the fold limbs, paleomagnetists began to realize that the internal strain of rocks could rotate a pre-folding

remanence to appear 'syn-folding' (Facer 1983; van der Pluijm 1987; Kodama 1988).

To understand the effects of rock deformation on the magnetization of a sedimentary rock, we need to understand what type of sedimentary rocks paleomagnetists typically measure and how those rocks will be deformed in an orogenic belt. Paleomagnetists have obtained the best sedimentary rock results from fine-grained clastic rocks and from carbonates. The fine grain size of the clastics ensures a quiet depositional environment that allowed the magnetization of the magnetic particles to align easily with the ambient geomagnetic field during settling. Fine-grained rocks, however, tend to soak up the tectonic strain in an orogenic belt more easily because they are usually less competent. Fine-grained clastic rocks are therefore more susceptible to any rotation of paleomagnetic remanence or strain-induced remagnetization that might occur. Carbonate rocks are important rock types in orogenic belts and usually sampled by paleomagnetists. The previous chapter (Chapter 6) showed however that both of these rock types were affected by

---

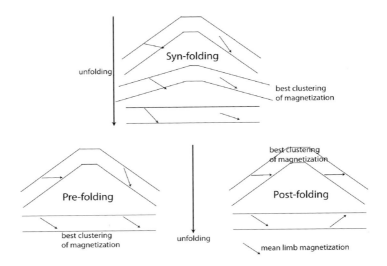

Fig. 7.1 Pre-folding, syn-folding and post-folding magnetization geometries.

an orogenically driven fluid event in the Appalachians during the Late Paleozoic (Oliver 1986; McCabe & Elmore 1989). The clastic rocks were partially remagnetized and the carbonate rocks were completely remagnetized, so paleomagnetists approach carbonate rocks in mountain belts with a good deal of apprehension and suspicion (or at least they should).

The types of strain that would be expected for the rocks typically sampled by paleomagnetists in mountain belts are those that result from relatively low-pressure and low-temperature conditions. As rocks are exposed to higher temperatures and pressures that would cause metamorphism at greenschist grades or higher, magnetizations are reset either thermally or by the growth of secondary magnetic minerals; these high-grade rocks are therefore usually avoided by paleomagnetists. The grain-scale deformation mechanisms of the low-grade rocks, particularly folded rocks, are spaced cleavage (axial planar) formation, pressure solution and particulate flow (i.e. grain boundary sliding). Flexural slip/flow and tangential-longitudinal strain geometries are particularly important for the low-grade folded rocks sampled by paleomagnetists. The deformation of folded rocks is ultimately the result of pure and simple shear strain. Pure shear strain causes a complicated vector displacement field (Ramsay & Huber 1983) but results in co-axial strain (Fig. 7.2). The strain ellipse created by pure shear strain does not

change the orientation of its axes as strain increases (Fig. 7.2). Simple shear strain has a relatively simple vector displacement field (Ramsay & Huber 1983), but causes non-coaxial strain. The axes of the strain ellipse created by simple shear rotate as the amount of strain increases.

Flexural slip/flow is the strain most associated with folded sedimentary rocks and should be considered by paleomagnetists whenever they work in deformed rocks. Flexural slip and flexural flow are the end members of a continuum of behavior. In flexural slip the strain is discontinuous at the grain scale and is concentrated at slip surfaces in the fold, usually bedding planes (Fig. 7.3); in flexural flow however the strain is distributed continuously throughout the rock. The observation of slickensides on bedding plane surfaces in folded rocks is good evidence that flexural slip has occurred. If some of the simple shear strain that caused the flexural slip was also distributed through the rock, then flexural flow also occurred. It is difficult to know, without making strain measurements, how these different types of strain were partitioned in the rock. Flexural slip/flow is caused by bedding-parallel simple shear so it is non-coaxial strain and is the type of strain that could cause problems for the interpretation of the fold test. Facer (1983) points out that concentric or parallel folds in which bed thickness stays the same around the fold are typically affected by flexural

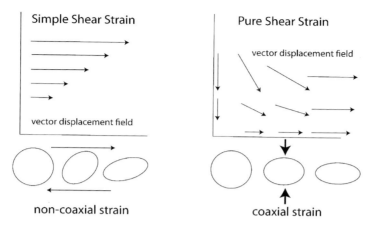

Fig. 7.2 Vector displacement fields for simple shear and pure shear strain. Modified from Ramsay & Huber (1983). Strain ellipses showing rotation of strain ellipse for non-coaxial strain and no rotation of strain ellipse for coaxial strain.

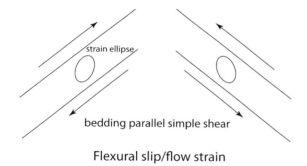

Fig. 7.3 Simple shear strain parallel to bedding results from flexural slip/flow folding. Strain is distributed continuously through the bedding for flexural flow folding. Strain is concentrated at the bedding planes for flexural slip folding. Arrows show the sense of shear parallel to the bedding of the fold limbs. The orientations of the strain ellipses on each fold limb are the result of the bedding-parallel shear non-coaxial strain.

flow/slip strain. Similar folds, in which the apparent thickness of a bed parallels the trace of the fold's axial plane and remains constant, can however have simple shear parallel to the fold's axial plane and affect the magnetization of a rock similarly to flexural flow/slip (Facer 1983).

The development of axial planar spaced cleavage at a high angle to bedding in a fold should also be impor-

tant for paleomagnetists to consider as they interpret their fold test results. Spaced cleavage is caused by two processes: the preferred alignment of platy minerals (usually phyllosilicates) either by physical rotation or growth in a differential stress regime, and the dissolution and removal of material by pressure solution from the cleavage domains. In our discussion, cleavage is formed by coaxial strain (pure shear) with the axis of maximum shortening perpendicular to the cleavage plane. The removal of material from the cleavage domains can cause an apparent oversteepening of the fold limb dips. The volume loss and resulting oversteepening can be important to the paleomagnetic fold test.

Finally, for the very low and low metamorphic grade rocks in mountain belts typically measured by paleomagnetists, the grain-scale behavior of sedimentary rocks during these different kinds of strain is important. Borradaile (1981) has suggested that particulate flow (grain boundary sliding) in which the individual grains in a sedimentary rock do not deform but flow past each other as the rock deforms can be an important grain-scale deformation mechanism. Of course, Borradaile points out, this is simply one end member of a continuum of behavior. Particulate flow can be coupled with different amounts of individual grain deformation. For particulate flow without any grain shape changes however, the standard structural geology techniques of measuring bulk rock strain, i.e. Fry center-to-center measurements or R_f-ϕ measurements, will not record any rock strain. This is important to realize when documenting bulk strain for interpreting the fold test of the low-grade rocks. Particulate flow is what van der Pluijm (1987) considers to be rigid body rotation of rock grains in his discussion of the effect of internal deformation of rocks on the paleomagnetic fold test. Van der Pluijm (1987) considers the other end member of grain-scale deformation to be homogeneous strain in which the grains of the rock change their shape. This type of rock strain can be easily documented by Fry and R_f-ϕ strain measurement techniques.

Whether the non-magnetic grains of the rock and the magnetic mineral grains that carry the remanence of the rock deform by particulate flow or by homogeneous strain becomes the focus of attention when considering the effects of tectonic strain on remanence. If paleomagnetic remanence rotates due to rock strain, and this itself is a matter of contention, it is important to understand how the magnetic grains behave. If they behave as rigid particles and rotate due to simple shear, the remanence can rotate through the shear plane following the equations of Jeffery (1923; Fig. 7.4). If the magnetic particles passively follow the deformation of the homogeneously deforming rock particles, the remanence will rotate according to the equations of March (1932) in what is called March or passive marker strain. The magnetization will behave like a line drawn on the deforming rock body, and will not be able to rotate through the shear plane during simple

shear. The behavior of the magnetic mineral grains during deformation, rigid particle versus passive marker, will be very important to how the paleomagnetic fold test is interpreted if internal deformation of the rock has affected the remanence as well as the rigid body rotation of the fold limbs. If the remanence can rotate through the shear plane, which is the bedding plane in flexural flow/slip folding, then a syn-folding configuration of the magnetization can be created from an initial pre-folding magnetization (van der Pluijm 1987; Kodama 1988). The evidence (both field observations and laboratory experiments) for passive marker versus rigid particle rotation was the focus of remanence rotation due to tectonic strain studies in the 1980s and 1990s (Kligfield *et al.* 1983; Hirt *et al.* 1986; Kodama & Goldstein 1991; Cogne & Perroud 1985; Stamatakos & Kodama 1991a, b; Borradaile 1993); a satisfactory resolution has yet to be obtained.

It is instructive to consider the sequence of deformation that sedimentary rocks typically experience in the evolution of an orogenic belt, as the accumulated strains could have an effect on the rock's remanence and should be considered, in light of possible remanence rotation due to rock strain, in interpreting a rock's paleomagnetism. As an illustration, we will use Nickelsen (1979) and Gray & Mitra (1993) findings for the Appalachian fold belt. Other orogenic belts can have similar strain histories, but may differ in some

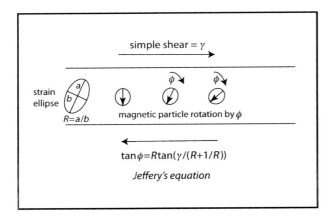

Rigid Particle Rotation

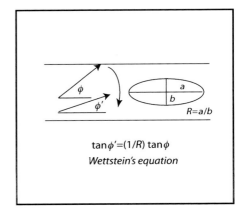

Passive Marker Rotation

Fig. 7.4 Rigid particle versus passive marker rotation of paleomagnetic remanence.

detail (see, for example, Yonkee & Weil's 2010 study of strain in the Wyoming salient). In the Appalachians, Alleghanian-age deformation caused a two-tier thrust system. The lower thrust sheet is composed of Cambro-Ordovician carbonate rocks and flysch while the upper thrust sheet comprises Silurian to Pennsylvanian silici-clastic rocks (Gray & Stamatakos 1997). The lower sheet is deformed primarily by layer-parallel shorten-ing and large-scale thrust faulting while the upper thrust sheet has undergone a sequence of deformation that includes first layer-parallel shortening and top to the foreland layer-parallel shear, then flexural slip/flexural flow folding and finally fold modification and late-stage thrust faulting. The Wyoming salient (Yonkee & Weil 2010) was also dominated by early layer-parallel shear, but then suffered from a sequence of tangential extension and layer-parallel shear. In both salients, layer-parallel shortening caused by a horizontal shortening axis has the potential to steepen the paleomagnetic inclination while fold modification in the Appalachians caused by axial planar cleavage or tightening of folds by homogeneous strain could also potentially rotate the remanence into the cleavage plane (Cioppa & Kodama 2003b). The effects of flexu-ral slip/flow (i.e. layer-parallel shear) are considered in detail in the following section.

REMANENCE ROTATION AND INTERNAL DEFORMATION OF ROCKS DURING FOLDING

In a theoretical approach, van der Pluijm (1987) con-sidered the effects of rigid particle and homogeneous bulk strain during folding when rocks are deformed by flexural flow/slip strain. He showed that a pre-folding magnetization could be rotated at the grain scale to appear to be syn-folding. In his consideration of the effects of internal strain on the paleomagnetic fold test, Facer (1983) did not explicitly show that a syn-folding magnetization could result from folding strain, but did point out that flexural slip/flow strain and shear paral-lel to a fold's axial plane would have essentially the same effects on a rock's magnetization.

Using numerical modeling, Kodama (1988) consid-ered in detail the effects of flexural slip/flow strain and tangential-longitudinal strain during folding on a rock's paleomagnetic remanence. For flexural slip/flow deformation, he considered a fold with 45° limb dips and an initial Fisher distribution of paleomagnetic

directions. Based on Ramsay (1967), the amount of simple shear parallel to the fold's bedding planes was equal in magnitude to the limb dip in radians. Kodama considered the effects of flexural slip/flow strain on the magnetization for two different grain-scale strain mechanisms: if the magnetization rotated as a passive line or if it rotated as if it were carried by rigid, actively rotating magnetic particles. For the passive line rota-tion he used the equations of March (1932) to deter-mine the orientation and magnitude of the strain ellipse that would be created in each limb due to flexu-ral flow strain (continuously distributed through the bedding) and used Wettstein's equation (Ramsay & Huber 1983) to calculate how much each paleomag-netic direction would be rotated toward the long axis of the strain ellipse (Fig. 7.4). For rigid particle rotation he considered that each paleomagnetic direction of the Fisher distribution was the resultant of individually magnetized particles that rotated like ball bearings in bedding-parallel simple shear using the equations of Jeffery (1923). Both equi-dimensional and prolate magnetic particles were modeled. After the magnetiza-tions were rotated by either passive line marker or rigid particle strain, the deformed Fisher distributions were stepwise unfolded to determine at what stage of partial unfolding the best clustering of magnetic directions occurred. The significance of the clustering was checked using the statistical test of McFadden & Jones (1981).

Simple consideration of the effects of rigid particle rotation due to flexural flow/slip bedding-parallel simple shear shows that the sense of particle rotation is opposite to that of the rigid body rotation of the fold's limbs (Fig. 7.5). In a sense, the grain-scale strain 'undoes' the effects of rigid body rotation of the fold's limbs, and will always lead to the rotation of a pre-folding magnetization to appear to be syn-folding. The numerical modeling supports this intuitive under-standing and shows that rigid particle rotation under the effects of flexural flow/slip strain will always lead to the best clustering of paleomagnetic directions at about 50% unfolding, no matter what the inclination of the initial paleomagnetic direction is. This result assumes that the magnitude of simple shear strain is equal to the limb dip in radians.

If the magnetization rotates as a passive line marker due to flexural flow/slip strain, the results are entirely different. This is because during passive line marker strain the magnetization cannot rotate through the shear plane, i.e. the bedding plane in a flexural flow

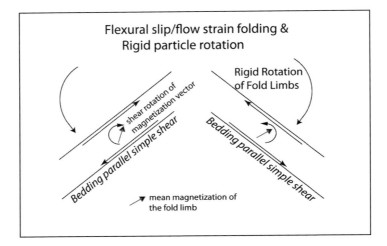

Fig. 7.5 The effect of flexural slip/flow folding strain on a magnetization if it rotates as if carried by rigid particles by Jeffery's (1923) equation. Note that the sense of magnetization rotation is opposite to the rigid body rotation of the fold limbs, causing a pre-folding magnetization to be rotated into a syn-folding configuration.

fold; only very steep-inclination initial paleomagnetic directions will be rotated to appear syn-folding (Fig. 7.6). The mechanism by which magnetizations rotate at the grain scale, either passive line or rigid particle, becomes extremely important in understanding the effects of flexural slip/flow strain on interpretation of the fold test.

The bulk flattening of a fold by homogeneous strain is a special case of remanence rotation as a passive marker in flexural flow strain, because the vertical strain ellipses that result from the homogeneous flattening are similar in orientation to those that arise from flexural flow.

An entirely different folding strain configuration that could affect the magnetization of a fold is tangential-longitudinal strain. In this case, simple shear does not occur during folding. Therefore, only passive line marker rotation of the remanence was considered by Kodama (1988). In tangential-longitudinal strain the outer arc of the fold undergoes bedding-parallel extension and the inner arc of the fold undergoes bedding-parallel compression (Fig. 7.7). Using numerical modeling, Kodama considered the effects of this strain when one limb was sampled under the neutral surface in the bedding-parallel compression regime and the other limb was sampled above the neutral surface in the bedding-parallel extension regime. He also considered the case where both limbs

were sampled above and below the neutral surfaces. Magnetizations were rotated into the extensional axes of the strain ellipses by Wettstein's equation. Only in the special case of sampling one limb above and one limb below the neutral surfaces could a syn-folding magnetization result from a pre-folding magnetization. Fold curvatures of approximately 40% for initially intermediate inclinations were needed for the most pronounced syn-folding magnetization configurations.

Finally, Kodama (1988) considered the effects of discontinuous deformation during folding: the formation of axial planar cleavage. The volume loss in the cleavage domains leads to an oversteepening of the fold limbs with respect to the bedding planes within the undeformed microlithons between the cleavage planes. If the oversteepened limbs are used to untilt the paleomagnetic directions in each limb, a syn-folding magnetization can result from a pre-folding magnetization. Kodama's modeling found that a volume loss of about 50% for limb dips between 30 and 60° could lead to a maximum clustering of directions at about 50% unfolding. Smaller volume losses will lead to best clustering at partial unfoldings closer to 100%. If the best clustering occurs at partial unfoldings of 85% or greater, the magnetization will not be distinguishable from pre-folding using the McFadden & Jones (1981) statistical test.

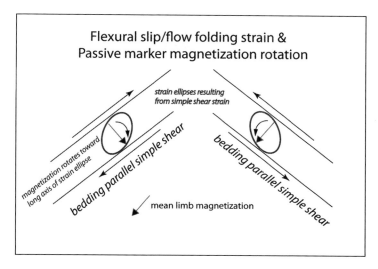

Fig. 7.6 The effects of flexural slip/flow folding strain when the magnetization rotates as a passive marker and the initial inclination is very steep in bedding coordinates. In this special case the magnetization rotating into the long axis of the strain ellipse is in the same sense as a rigid particle magnetization rotating in the bedding-parallel simple shear strain. Since the magnetization would rotate opposite to the rigid body rotation of the fold limbs (same sense of limb rigid body rotation as in Fig. 7.5) a syn-folding configuration would result. This would not be true for an initially flat magnetization in bedding coordinates. (See Colour Plate 11)

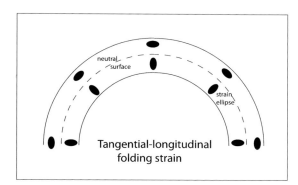

Fig. 7.7 Strain configuration for tangential-longitudinal strain during folding. Since no simple shear occurs, magnetizations would rotate into the long axes of the strain ellipses by passive marker rotation.

REMANENCE ROTATION IN THE FIELD

Theoretical work of Facer (1983), van der Pluijm (1987) and Kodama (1988) on the effects of folding strain on a rock's magnetization showed all the possibilities for different strain geometries and different grain-scale rotation mechanisms. For a paleomagnetist, the most important internal strain to consider turns out to be the commonly observed folding strain of flexural slip/flow. Many folds show slickensides on the bedding planes of their limbs so, at the very least, flexural slip is occurring in these folds. What Kodama's numerical modeling showed is that the grain-scale deformation of the paleomagnetic particles, either rigid particle rotation or passive line marker rotation, was critical in determining whether a pre-folding remanence was rotated into a syn-folding configuration by flexural slip/flow strain. Workers therefore looked to evidence from deformed rocks and from laboratory experiments to help understand how remanence rotates during deformation.

As part of his examination of folding strain in remanence, Kodama (1988) also looked at rock strain data from two folds – one in carbonate rocks and one in siliciclastic rocks – to determine if strain could have caused syn-folding behavior and to see how a magnetization deformed at the grain scale. In the Cambrian dolomite of the Allentown Formation, the magnetization in a small fold from a quarry in downtown Bethlehem, Pennsylvania, had already been shown to be

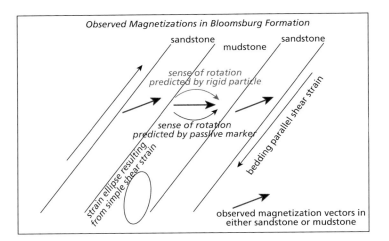

Fig. 7.8 Schematic representation of remanence rotation experiment in the Bloomsburg Formation red beds (Kodama 1988). The mudstone magnetization was rotated with respect to the magnetizations in the surrounding sandstone beds. The relative sense of rotation suggests rigid particle behavior since it is opposite to that predicted by passive line marker behavior of rotation into the long axis of the strain ellipse that would result from flexural slip/flow strain. (See Colour Plate 12)

syn-folding in configuration (Stead & Kodama 1984). Kodama (1988) measured the strain in the fold limbs using the R_f-ϕ technique (Ramsay & Huber 1983). The most important results of this study are as follows.
1. The amount of strain was not the same in both limbs; it was greater and oriented in the sense expected for flexural flow folding in one limb, but smaller and not in the flexural flow sense for the other limb.
2. The maximum magnitude of strain measured was only about one-quarter of that predicted by Ramsay's theoretical prediction that simple shear would equal limb dip (in radians).

Based on modeling, the amount of strain in the Allentown Formation fold was not enough to rotate the remanence of these carbonate rocks into a syn-folding configuration. To understand the effects of folding strain in siliciclastic rocks, the Silurian Bloomsburg Formation red beds were studied in a fold near Palmerton, Pennsylvania. Only one limb (the steeper dipping southern limb) of the syncline was exposed so the study consisted of comparing the remanence in two different lithologies in the limb – two sandstone beds that stratigraphically bracketed a mudstone bed – with the assumption that the less-competent mudstone bed would soak up more strain than the sandstone beds. No strain was measured in the beds; only their remanences were compared. The mudstone mean paleo-

magnetic direction was observed to be 10° more easterly than the sandstone mean paleomagnetic direction. If the mudstone experienced more flexural flow bedding-parallel shear strain than the bracketing sandstone beds, its magnetization should have been more westerly (Fig. 7.8) than the magnetization of the sandstone beds for passive line marker strain and more easterly for rigid particle rotation. The results therefore support rigid particle rotation for a paleomagnetic remanence carried by the hematite particles of the Bloomsburg Formation red beds. Rigid particle modeling of the 10° offset between the mudstone and sandstone magnetizations further indicates that the mudstone could have experienced only about one-third of the predicted maximum shear strain based on the bedding dip. For both carbonates and siliciclastic rocks, the amount of strain experienced is quite a bit less than the maximum predicted by Ramsay (1967) based on the limb dip; rigid particle motion is therefore suggested as the grain-scale mechanism for remanence rotation.

Rigid particle rotation of remanence was also observed in detailed studies of strain and remanence for the Bloomsburg Formation and the Mississippian Mauch Chunk Formation red beds (Stamatakos & Kodama 1991a, b). Stamatakos and Kodama studied the strain and remanence in the Bloomsburg Forma-

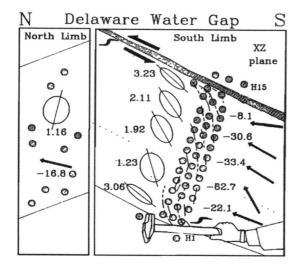

Fig. 7.9 Detailed finite strain measurements at a fold in the Bloomsburg Formation at Delaware Water Gap shows strain ellipse orientations consistent with flexural flow folding strain at the southern limb of the fold (Stamatakos & Kodama 1991b). Magnetization vectors measured in detail throughout the fold limb show orientations consistent with rigid particle rotation because the magnetization exposed to the greatest strain (strain ellipse 3.23) has rotated through the bedding plane (shear plane). J Stamatakos and KP Kodama, The effects of grain-scale deformation on the Bloomsburg Formation pole, *Journal of Geophysical Research*, 96, B11, 17919–17933. Copyright 1991 American Geophysical Union. Reproduced by permission of American Geophysical Union.

tion for three folds in Pennsylvania and found that the inclination of the remanence was directly correlated to the strain, measured by the Fry (1979) and R_f-ϕ center-to-center techniques, and rotated in the sense predicted by rigid particle rotation due to the bedding-parallel simple shear caused by flexural flow (Fig. 7.9). For the Mauch Chunk Formation, Stamatakos & Kodama (1991a) studied one first-order fold in Lavelle, Pennsylvania again measuring the strain with the center-to-center techniques to compare it to the remanence directions in the fold. The strain in the fold was consistent with flexural flow strain and concentrated in the finer-grained lithologies, and not in the more competent coarser sandstone layers as expected. For the mudstones there was a direct correlation between the paleomagnetic inclination and the magnitude of strain, and the sense of rotation for the inclina-

tion was consistent with rigid particle rotation of the hematite magnetic particles.

The R_f-ϕ and the normalized Fry center-to-center structural geology techniques were used to quantify strain in carbonates and siliciclastics because the magnetic measure of strain by anisotropy of magnetic susceptibility (AMS) has been found to be inadequate. AMS is a complicated signal arising from both paramagnetic grains and ferromagnetic grains that react to strain by different degrees, so a straightforward correlation between AMS and strain is not possible (Borradaile & Jackson 2004). The R_f-ϕ technique measures strain from the ellipticity and orientation of initially elliptical grains that are reoriented and further deformed by strain (e.g. Ramsay & Huber 1983). The normalized Fry center-to-center technique uses measurement of the distances between the centers of neighboring quartz grains in a siliciclastic rock. The deformation and reorganization of the quartz grains is a measure of bulk strain in the rock.

REMANENCE ROTATION IN THE LABORATORY

Realizing the importance of determining whether a paleomagnetic remanence deformed as rigid particles or as a passive line during the bedding-parallel simple shear strain during flexural flow/slip folding, Kodama & Goldstein (1991) conducted simple shear box experiments with mixtures of silicone putty and magnetite or kaolinite clay and magnetite. The mixtures were magnetized by applying an ARM at high, moderate and low angles to the shear plane and then deformed. Twelve out of twenty runs unequivocally contradict passive line behavior for the ARM. Modeling of the results is in strong agreement with rigid particle rotation of the magnetite grains carrying the remanence.

Graham Borradaile has also done extensive work to determine how paleomagnetic remanence rotates during rock deformation. Borradaile's laboratory strain work (summarized in Borradaile 1997) provided an entirely different perspective and results from those found by Kodama and colleagues. Borradaile used a computer-controlled triaxial pressure rig for his experiments that imposed high confining pressures ($\geq 200\,MPa$) and low differential stresses ($\leq 100\,MPa$) at low strain rates (10^{-5}–$10^{-6}\,sec^{-1}$). For his rock analogues, Borradaile used a mixture of either magnetite or hematite as the magnetic mineralogy in a non-magnetic matrix of calcite bonded with Portland cement. Borradaile achieved strains of up to 45% shortening.

For simple coaxial strain with up to 45% oblate shortening and one component of magnetization, Borradaile (1993) observed simple passive line marker behavior for the rotation of remanence with the magnetization rotating away from the shortening axis by amounts predicted from the strain of the rock analogue and Wettstein's equation. It is important to realize that Borradaile's experiments did not examine the effects of non-coaxial strain, as might be expected for flexural slip/flow folding. Non-coaxial strain was what Kodama observed to cause rigid particle behavior of the remanence-carrying magnetic grains.

For multi-component magnetizations the results of coaxial strain were, in some cases, counter-intuitive. For these experiments two perpendicular components of magnetization were applied to the samples, either using an ARM or an IRM. One component was high coercivity and the other component low coercivity. For the magnetite-bearing samples, strain caused a hardening (increase) of the overall coercivity of the samples reducing, preferentially, the low-coercivity component of magnetization by presumably affecting the large multi-domain magnetite grains. The magnetization was observed to rotate into the direction of the initially harder component of magnetization with no regard to the direction of the shortening axis. If the initial harder component of magnetization was parallel to the shortening axis, the magnetization was observed to rotate into the shortening axis; this was not a result expected for passive line rotation or rigid particle rotation of the remanence. For hematite-bearing samples with multi-component magnetizations the results were exactly the opposite. This time the magnetization rotated into the softer of the two imposed components of magnetization, again despite the direction of shortening. In this case Borradaile interpreted the result to mean that the fine-grained high-coercivity hematite particles were dispersed by the strain. Finally, Borradaile (1994, 1996) deformed some samples carrying a single component of magnetization by strains of only 15%. These samples were observed to pick up a soft magnetization parallel to the 30 μT field in the triaxial rig, providing an example of strain-induced remagnetization i.e. a piezo-remanent magnetization.

Borradaile's work shows that the effects of rock strain on remanence can be very complicated partly due to the grain-scale effects of strain on the individual magnetic particles either changing their magnetic domain state, rotating particles heterogeneously or causing their remagnetization. As he points out in his review article (Borradaile 1997), his results should be extended to the natural world with caution because the strain rates in the laboratory were 10^6 times greater than those found in nature.

FURTHER FIELD AND LABORATORY OBSERVATIONS

There has been little subsequent work on remanence rotation due to tectonic strain, but the work done has not supported rigid particle rotation. However, the geometry of the strain related to the remanence has not been particularly conducive to unraveling the effects of passive line versus rigid particle behavior. Kirker & McClelland (1997) looked at the deflection of remanence during progressive cleavage development in kink bands in the Devonian Pembrokeshire red sandstones of southwest Wales. Remanence was deflected related to the amount of kinking. Some samples' magnetizations were rotated through local bedding, a sign of rigid particle behavior, but most were not. Since the magnetization is clearly syn-folding, the magnetizations were not exposed to as much strain as they would have been if they were pre-folding in age. Cioppa & Kodama (2003b) looked at strain and remanence in a zone of intensified axial planar spaced cleavage in the red beds of the Mississippian Mauch Chunk Formation. Strain intensity was measured with center-to-center techniques. The magnetization was observed to rotate into the axial planar cleavage developed in the rocks as the amount of strain increased, but the pure-shear coaxial strain did not allow the distinction between passive marker and rigid particle behavior.

Till et al. (2010) conducted a high-pressure (300 MPa) high-temperature (500°C) deformation experiment to see the effects of greenschist grade metamorphism on magnetic remanence and magnetic anisotropy. The deformation involved simple shear along a plane perpendicular to the weak thermal remanence applied to the samples; however, the shear was induced by coaxial stress at 45° to the shear plane. Some of the samples were created with a weak initial fabric, and other samples without any initial fabric. The effects on magnetic anisotropy were the main focus of the experiment and the authors found that the samples without initial fabrics developed magnetic fabrics during deformation that could detect increasing levels of strain, while samples with initial fabrics could not. Pertinent to our discussion here, the post-deformation remanence of the initially fabric-less samples was interpreted by the authors to be a remagnetization based on a change in magnetic properties, i.e. a piezo-remanent magnetization. The remagnetization was parallel in direction to the pre-deformation remanence for these samples. The samples with initial fabrics had a post-deformation remanence that was either parallel to the pre-deformation remanence or scattered. Unfortunately, these results do not inform our understanding of how remanence deforms during non-coaxial simple shear strain, but they do show that the initial strain state of a sample will affect how well magnetic anisotropy can measure subsequent deformation. These results point to the complexity of using magnetic fabrics to measure the degree of rock deformation. Borradaile & Jackson (2004) detail these complexities and conclude that AMS is not a reliable way to quantify rock strain; however, it can determine the orientation of the strain ellipse.

STRAIN REMAGNETIZATION

The Till *et al.* (2010) study does show that tectonic strain can likely lead to remagnetization of a rock. Hudson *et al.* (1989) argue that a piezo-remanent magnetization has remagnetized the folded Preuss Formation sandstone in a Wyoming thrust sheet. Measurements of strain in the folded rocks suggest that the amount and geometry of the strain could not have caused over-steepening of the fold limbs, grain boundary sliding or flexural flow strain. Although the magnetization of the Preuss is syn-folding in geometry, detrital magnetite grains carry it. There is no evidence of secondary magnetic grains that could carry a CRM. Furthermore, the thermal history of the rocks does not point to a viscous partial thermal remanence as a cause of remagnetization. The authors are left with a stress-induced cause of remagnetization: a piezo-remanence. Housen *et al.* (1993) appeal to an entirely different mechanism for a strain-caused remagnetization of the Ordovician Martinsburg Formation shales and slates in eastern Pennsylvania. In a detailed sampling of a shale–slate transition in the Martinsburg Formation, Housen *et al.* observe dissolution and neocrystallization of magnetite leading to a CRM remagnetization mechanism caused by increasing strain and coincident formation of a spaced slaty cleavage in the rocks.

Strain-induced remagnetization was the conclusion of two detailed strain–remanence studies of first-order folds in the Appalachian mountains of West Virginia. Lewchuk *et al.* (2003) studied the Patterson Creek anticline and Elmore *et al.* (2006) studied the Wills Mountain anticline. Both studies investigated the Silurian Tonoloway Formation and Ordovician Helderberg Formation carbonate rocks that were folded in both anticlines and carried a Late Paleozoic remagnetization. The pattern was basically the same in both studies; the Tonoloway Formation rocks had either a pre-folding (Patterson Creek anticline) or early syn-folding (Wills Mountain anticline) magnetization, while the Helderberg Formation had a true syn-folding magnetization in the folds. Magnetite is the magnetic mineral responsible for the magnetizations. Detailed measurements of strain in the rocks showed that the syn-folding magnetizations were in rocks with more pressure solution strain, either compactional or tectonic in geometry. The authors of the studies argue for a modification of the rocks' magnetization caused by strain or a strain-induced remagnetization modified by a piezo-remanent magnetization. Mechanical rotation of pre-folding magnetic grains by strain is ruled out in the Wills Mountain anticline study because the difference in the amount of strain between the pre-folded and syn-folded rocks is not enough to explain the syn-folding geometry of magnetization.

OROCLINES AND PALEOMAGNETISM

Paleomagnetism is one of the best ways of studying how oroclines form. Oroclines are the bends observed in many mountain chains globally. Weil & Sussman (2004) show 11 major oroclines throughout the world. The Pennsylvania salient in the Appalachians is a mountain chain bend that has been studied the most, both by paleomagnetic and structural geology methods. Pertinent to our discussion of strain and paleomagnetism, grain-scale strain was considered at one time to be the possible cause of divergent declinations around the Pennsylvania salient.

As Weil & Sussman (2004) point out there is a long history to the idea of an orocline, the bending of a once-straight fold belt. The name orocline was first proposed by Carey (1955), and ideas about the formation of oroclines have evolved since then. Weil & Sussman

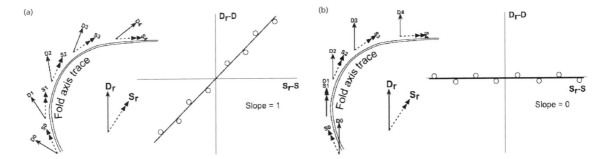

Fig. 7.10 The paleomagnetic declination test for oroclinal bending (from Weil & Sussman 2004): (a) a true orocline in which the bend of the mountain belt is revealed by paleomagnetic declination and (b) a primary arc in which there is no vertical axis rotation of declination around the fold belt. A progressive arc would have declinations around the arc intermediate between the two cases.

(2004) summarize the current thinking that there are three kinds of bends in mountain chains as follows.

1. Primary arcs have been curved in map view from the beginning. They formed in a curve and the curve has not tightened during deformation.

2. At the other end of the spectrum is the true orocline that formed as a linear mountain belt and was subsequently bent into a curved shape during deformation.

3. In between the two end members are progressive arcs that bend as they are forming; they acquire their curvature during deformation of the mountain belt.

The tests needed to distinguish between these different types of mountain chain bends depend primarily on measuring the changes in paleomagnetic declination around the curved mountain belt. Plots of declination with respect to the strike of the bedding around the curve should show a linear relationship if the mountain belt is a true orocline (Fig. 7.10). The deflection of the paleomagnetic declination should follow the deflection of the bedding strike. For progressive arcs, the declination will be less rotated than the strike. Primary arcs would not show any declination rotation around the arc of the belt. The Cantabrian arc in Spain is a good example of a true orocline (Weil *et al.* 2001). Plots of the magnetic declination versus strike show evidence that 100% of the strike variations are due to true oroclinal bending.

The results for the Pennsylvania salient, one of the most-studied arcs, are much more complicated. The main finding about the Pennsylvania salient revealed by paleomagnetism is that secondary magnetizations in the folded rocks around the arc show absolutely no difference in declination. Many of these secondary

magnetizations are syn-folding in age and were acquired during the large fluid-driven Kiaman age remagnetization event observed throughout the Paleozoic rocks of the Appalachians. This observation is critical since it means that once the folds were forming the bend in the Pennsylvania salient already existed, the folds formed in a bent geometry (Stamatakos & Hirt 1994; Stamatakos *et al.* 1996). However, the prefolding components of magnetization of the red beds of the Silurian Bloomsburg and the Mississippian Mauch Chunk formations do show a declination difference around the salient, with the directions in the northern limb of the arc pointing north and the directions in the southern limb pointing northwesterly (Kent & Opdyke 1985; Miller & Kent 1986; Kent 1988).

Initially, Stamatakos & Kodama (1991b) argued that the divergence in declinations could be the result of grain-scale strain rotating remanence by thrusting to the north in the northern limb of the salient and thrusting to the northwest in the southern limb of the salient. This would predict that inclinations would shallow but stay northerly in the northern limb of the salient due to top-to-the-north bedding-parallel shear. It would also predict that inclinations would shallow and declinations would move northwesterly in the southern limb of the salient due to top-to-the-northwest bedding-parallel shear. Stamatakos & Kodama (1991b) saw this pattern at two sites, one from the northern limb at Montour Ridge, Pennsylvania and one from the southern limb at Round Top, Maryland. Of course, they also observed rigid particle rotation of remanence within first-order folds around the salient that rotated a pre-folding remanence to appear syn-folding and this informed their interpretation of diverging declinations around the curved mountain belt. However, when Stamatakos & Hirt (1994) plotted all the pre-folding magnetizations from many different localities from the southern limb of the salient, they did not observe the predicted pattern of shallower inclinations being coincident with more northwesterly declinations if grain-scale strain were an important factor. One possibility is that grain-scale strain did in fact affect some of the remanences, as shown in the detailed studies of the folds, but is not widespread enough to be used as an explanation for the declination divergence between the limbs of the salient.

Gray & Stamatakos (1997) used these results to construct a model for the formation of the Pennsylvania salient that was compatible with both paleomagnetic data and structural geology data. They argue that the rocks deformed as two separate sheets: the lower Cambro-Ordovician carbonates and the upper Siluro-Devonian siliciclastics. Laterally varying amounts of layer-parallel shortening in the lower sheet caused vertical axis block rotations in the upper sheet recorded by the pre-folding magnetizations in the siliciclastics. Imbrication by thrusting in the lower sheet caused a thickening toward the middle of the salient and divergent shortening directions and folding in the upper sheet due to gravitational spreading.

Cederquist *et al.* (2006) tested this model by measuring the paleomagnetism of the Cambro-Ordovician carbonates at 10 anticlines around the salient and found that all were remagnetized during the Late Paleozoic, as are many carbonates in the Appalachians. The main finding of this study is that all directions had essentially the same declination around the salient. Gray & Stamatakos's (1997) lower sheet showed no bending since the time when the folds were formed in it. Ong *et al.* (2007) took the testing of the Gray & Stamatakos (1997) model further not with paleomagnetic data but with calcite twinning data that measures the earliest strain in the rocks, in this case the layer-parallel shortening that occurs before buckling into folds. Ong *et al.* (2007) applied the same test used for paleomagnetic declinations around a curved mountain belt to the shortening directions obtained from calcite twinning and found evidence for an oroclinal bending of shortening directions around the salient for both the Cambro-Ordovician and the Siluro-Devonian

rocks, the two supposedly independently deforming sheets of Gray & Stamatakos (1997). However, most of Ong *et al.*'s data is from the southern limb of the salient, particularly for the Siluro-Devonian data for which there is only one site on the northern limb. If the Cambro-Ordovician and Siluro-Devonian data are considered separately instead of combined, then the Siluro-Devonian data has a consistently more southeasterly declinations (by 15°) than the Cambro-Ordovician rocks; this perhaps suggests a difference in the deformation of the two sheets. However, the calcite twinning data does suggest a simpler kinematic scenario for the formation of the Pennsylvania salient than that envisaged by Gray & Stamatakos (1997). This case study shows that understanding the bends in mountain chains requires more than paleomagnetic data alone.

Grain-scale strain rotation of remanence has not been considered as a significant factor in subsequent studies of oroclines. However, in their study of the Wyoming salient Weil *et al.* (2010) did briefly discuss the possibility of remanence rotation due to layer-parallel shortening observed in the rocks. High-strain sites were avoided in the paleomagnetic analysis of the salient, but strain–remanence relationships were not studied in detail.

CONCLUDING THOUGHTS

After a good deal of interest in the effects of grain-scale strain on the accuracy of paleomagnetic remanence in the 1980s and 1990s, little work has since been done on this important question. Work on the natural magnetizations of Paleozoic-age Appalachian folded rocks suggests that remanence may rotate at the grain scale as a rigid particle, but laboratory experiments and other field studies suggest that the importance of passive line marker rotation versus rigid particle behavior has not been fully resolved. Determining how paleomagnetic remanence rotates at the grain scale (if at all) during flexural flow/slip strain, a common strain mechanism during the folding of rocks typically sampled for paleomagnetism, will be critical for interpreting the paleomagnetic fold test as well as the accuracy of paleomagnetic directions.

Magnetization of Sediments and the Environment

The topic of environmental magnetism, really a sub-discipline of paleomagnetism and rock magnetism, is vast and diverse. The first comprehensive text on the field was written by Thompson & Oldfield (1986) which is fitting, given British researchers' pioneering efforts in the field. Since then several good books and review articles have been written on the field (e.g. Verosub & Roberts 1995; Maher 1999) and Tauxe's (2010) paleomagnetism textbook has devoted a chapter to the rock magnetic principles and theories that are applied to environmental issues. The most recent comprehensive coverage of environmental magnetism is the book by Evans & Heller (2003) that the reader should consult for a more complete review of the topic than is offered here.

This chapter will provide more specialized coverage and, after reviewing the basic rock magnetic principles and techniques used in environmental magnetic studies, concentrate on two topics: (1) the biogenic magnetic minerals or magnetosomes produced by magnetotactic bacteria and the use of their record in recent lake sediments as environmental indicators and (2) the recent application of rock magnetic parameters in stratigraphic sections of sedimentary

rocks to detect astronomically forced global climate cycles, thus providing high-resolution (nominally 10^4 yrs) time control.

ENVIRONMENTAL MAGNETISM TECHNIQUES: A BRIEF SUMMARY

In environmental magnetism studies, standard rock magnetic measurements are used to measure three different magnetic properties in sediments or sedimentary rocks: the concentration of magnetic minerals, the particle size of the magnetic minerals and the different types of magnetic minerals that are present in the sediment or rock. Two of these properties of the sediment, particle size and magnetic mineralogy, are determined primarily by using coercivity differences of different magnetic minerals or the different particle sizes of one magnetic mineral. Coercivity, as we have seen in earlier chapters, is a measure of a magnetic mineral's magnetic 'hardness' or the ability to withstand a change of its magnetization by the application of a magnetic field. The concentration of magnetic particles in a sample is measured by activating (mag-

Paleomagnetism of Sedimentary Rocks: Process and Interpretation, First Edition. Kenneth P. Kodama.
© 2012 Kenneth P. Kodama. Published 2012 by Blackwell Publishing Ltd.

netizing) different subpopulations of magnetic grains of different coercivities so that a measurement of the magnetized sample essentially 'counts' the number of magnetic grains, a measure of the amount of magnetic material. For the complete set of equations and basic rock magnetic principles the reader is referred to either Butler's (1992) or Tauxe's (2010) excellent paleomagnetism textbooks. We will provide only the most essential equations here and only a simple explanation of the rock magnetic principles; the reader is urged to read more deeply in Butler or Tauxe.

The two most common magnetic minerals in sediments are magnetite (Fe_3O_4) and hematite (Fe_2O_3). These two Fe oxides differ markedly in their crystal structure with magnetite being a cubic, inverse spinel and hematite being rhombohedral with a corundum structure and a hexagonal unit cell. These differences affect and control their ferromagnetism. The sub-lattices of the magnetite inverse spinel crystal have unequal numbers of antiparallel magnetic moments, thus making magnetite a ferrimagnetic mineral (as opposed to ferromagnetic which is the general term for minerals that have a permanent magnetism or remanence). The sub-atomic level magnetic moments that cause the spontaneous magnetization, or ferromagnetism, of the iron oxide minerals arises from interactions between the uncoupled electron spin moments in the d-shells of the iron atoms. The quantum mechanical interactions between the uncoupled spin moments of adjacent Fe atoms leads to a magnetic 'order' throughout the crystal. This magnetic structure leads to a strong spontaneous magnetization for magnetite (480 kA/m). Hematite, on the other hand, has a much weaker spontaneous magnetization (2 kA/m) because its crystal structure results in a canted antiferromagnetism. Essentially, equal numbers of antiparallel spin moments in the crystal are misaligned by only a very small angle (0.065°; Morrish 1994).

Because magnetite has a much stronger spontaneous magnetization than hematite, its particle-scale anisotropy (which controls the direction of magnetism in a particle) is dictated by the shape of the particle; in contrast, the particle-scale anisotropy of hematite is dictated mainly by its crystal structure. Magnetite particles tend to be magnetized along their longest shape axis, while hematite particles tend to be magnetized in the basal plane of the hexagonal crystal lattice. The different causes of particle anisotropy for these two magnetic minerals lead, in turn, to magnetite having much lower coercivities than hematite. Magnetite has

typical coercivities less than 100 mT while hematite's coercivities are several hundred to one thousand mT (Peters & Dekkers 2003).

Furthermore, the differences in spontaneous magnetizations between the two minerals cause different coercivity behavior for the particle sizes typically seen in natural sediments and sedimentary rocks (micron to submicron). Magnetite will be single domain (SD) (a particle is magnetized in the same direction throughout its volume) in the submicron range (c. 0.02–0.2 microns; Tauxe 2010) and multi-domain (MD) (a particle divides itself into volumes with different magnetic directions) at larger grain sizes. The subdivision into multi-domain grains occurs at a magnetic mineral's critical diameter, when it saves enough magnetostatic energy to build the walls between the different regions of different direction magnetization. By changing to a multi-domain structure, a magnetic particle minimizes its overall magnetostratic energy by arranging the north magnetic poles of one domain close to the south magnetic poles of an adjacent domain. Multi-domain particles can more easily adjust to externally applied magnetic fields by moving their domain walls to change the volumes of different domains in the grain. Domains parallel to the applied field become larger and those antiparallel become smaller. Magnetite's coercivity is then observed to increase as its grain size decreases into the SD grain-size range. Hematite, in contrast, will be single domain through much of its natural grain-size range. It switches to multi-domain configurations at particle sizes close to 10–20 microns (Dunlop & Ozdemir 1997).

Magnetic mineral concentration

A magnetic concentration measurement results from application of a laboratory remanence to a sample and then measurement of the strength of the sample's magnetization. The direction of the sample's magnetism is not important to this measurement; in fact, any natural remanent magnetization (NRM) of the sample is destroyed by the application of the laboratory field and the sample is no longer suitable for paleomagnetic study. As mentioned earlier, the laboratory-applied field 'activates' or magnetizes a subpopulation of the magnetic grains in a sample. Measurement by rock magnetometer essentially 'counts' the number of magnetic grains activated in the sample. The magnetic measurement is usually normalized by the mass of the sample, so the magnetic units typically employed to report the results are $A\,m^2/kg$.

Anhysteretic remanent magnetization (ARM) is used to quantify magnetite concentration or, more broadly, the concentration of low-coercivity magnetic minerals. One reason is that most laboratories can only apply ARMs with peak alternating fields up to 100–200 mT. One other environmentally important magnetic mineral in this class would be the iron sulfide greigite (Fe_3S_4) that is the sulfide analogue of magnetite. It has coercivities similar to that of magnetite because its crystal structure is identical with sulfur replacing oxygen and a spontaneous magnetization about one-quarter that of magnetite. Peters & Dekkers (2003) indicate that the iron sulfide pyrrhotite and the iron oxide maghemite also fall into the coercivity range that could potentially be activated by a 100 mT ARM but, as noted in Chapter 6, Horng & Roberts (2006) suggest that pyrrhotite is less likely to result from environmental processes such as reduction diagenesis. Maghemite (γ-Fe_2O_3) is important to environmental magnetic studies because it is commonly found in soils.

Isothermal remanent magnetization (IRM) is also used to measure magnetic mineral concentration in a sample. Since most laboratories can apply fields of at least 1 T and, in some cases, as high as 5 T, magnetic minerals with much higher coercivities can be activated by application of an IRM. Magnetization by an IRM is however a different process than that of ARM, and so magnetic minerals respond differently. ARM is considered a better analogue of thermal remanent magnetization (TRM) processes than IRM, but without the heating that could cause chemical changes of the magnetic minerals in a sample. TRM is really the 'gold standard' of activating the magnetization of particles in a sample because it is the process by which nearly all depositional magnetic grains were initially magnetized. The alternating field used to apply an ARM essentially 'loosens up' the magnetization of a magnetic grain so it can be reset by a weak magnetic field, similar in intensity to the Earth's magnetic field. This is the similarity to a TRM in which heating 'loosens up' the magnetization of a grain while the Earth's field (or a weak laboratory field) resets the grain's magnetization. IRM is more of a blunt force approach in which a DC magnetic field 'whacks' the magnetization of a grain into a new direction, depending on the orientation of the applied field with respect to the easy axis for the grain's magnetization and the grain's coercivity. A lightning strike is the natural process that remagnet-

izes a rock by an IRM. Nevertheless, IRM applied in the laboratory can be used to measure the concentration of magnetic minerals over a wide range of coercivities, so minerals such as hematite and goethite (an iron oxyhydroxide important in chemical weathering; α-FeOOH) as well as magnetite and iron sulfides can be activated. Goethite has coercivities up in the tens of Tesla (Peters & Dekkers 2003).

Finally, the last major magnetic measurement used to determine the concentration of magnetic minerals in a sample is susceptibility. Susceptibility measurements are different from the two remanence measurements mentioned above. Susceptibility is the induced magnetization that arises during the application of a low-strength magnetic field to a sample:

$$J_{ind} = \chi H$$

where J_{ind} is the induced magnetization, χ is the susceptibility and H is the applied field.

When the applied field is removed the sample loses its induced magnetization. The field strengths used for a susceptibility measurement are not large enough to give the sample a permanent magnetization like an IRM or an ARM.

One of the problems with measuring magnetic mineral concentration using susceptibility is that many different magnetic minerals have magnetic susceptibility, but due to different processes. Calcite and quartz are examples of common diamagnetic minerals that have weak negative susceptibilities, i.e. the induced magnetization is opposite to the applied magnetic field. Iron-containing silicates (e.g. micas, amphiboles, pyroxenes, clays) have paramagnetism with weak positive susceptibilities. Ferromagnetic minerals have strong positive susceptibilities. Usually the strong susceptibilities of the ferromagnetic minerals (magnetite and hematite) swamp the diamagnetic and paramagnetic contributions of silicates and calcite to the induced magnetization in a sample, but for some sedimentary rocks (e.g. limestones or 'clean' sandstones) the concentration of ferromagnetic minerals is so low that paramagnetic or diamagnetic minerals contribute significantly to the susceptibility.

In addition, the susceptibility of ferromagnetic minerals is a function of magnetic particle size, since the process of acquiring an induced magnetization is different for single-domain and multi-domain grains. Multi-domain grains readjust their domain structure to minimize their energy with respect to the applied field while single-domain grains can only rotate their remanence out of its easy direction of magnetization to respond to the applied field in a susceptibility measurement.

Multi-domain grains therefore have a stronger ferromagnetic susceptibility than single-domain grains. It's obvious that from this description of different susceptibility mechanisms that susceptibility is a complicated signal with many different possible sources in a sample. Even if the susceptibility of ferromagnetic minerals dominates a sample, the much larger spontaneous magnetization of magnetite with respect to hematite means that it will override the contribution of hematite to a sample's susceptibility, and make it hard to detect the hematite magnetically. Furthermore, it is not easy to deconstruct the susceptibility signal into its different sources. A remanence measurement can be deconstructed to a certain extent using the different coercivities of magnetic minerals and magnetic mineral grain sizes. This type of experiment is not possible for susceptibility. Susceptibility is still a valid method for measuring magnetic mineral concentration, but best used when one ferromagnetic mineral dominates the magnetic mineralogy or is the sole magnetic mineral of a sample.

Magnetic mineral grain size

Magnetic particle size, also called magnetic grain size, is measured by the ratios of different magnetic mineral parameters. Typical ratios include the ARM/χ ratio that quantifies the relative amount of fine-grained single-domain magnetite to the amount of coarse multi-domain magnetite. All bets are off if the magnetic mineralogy is contaminated by paramagnetics, diamagnetics or hematite. The ratio of the remanence measurements ARM/IRM can also be used to quantify magnetic particle size. ARM tends to respond more effectively to the finer-grained single-domain grains, particularly magnetite, while the IRM activates and responds to all magnetic grain sizes. This ratio should be used as a magnetic grain size measure when only one ferromagnetic mineral dominates the magnetic mineralogy of a sample.

Magnetic hysteresis parameters are used extensively to measure magnetic grain-size variations, but are more difficult to use on large suites of samples to monitor magnetic grain-size changes throughout a sedimentary section because each measurement takes a relatively long time (tens of minutes). In a hysteresis measurement, a sample is cycled through a DC field that changes its intensity and direction along one axis. The magnetization of the sample is measured while the magnetic field is being applied. A hysteresis loop results (Evans & Heller 2003, fig. 2.3) and four important parameters may be derived from the loop: the coercivity B_c, the saturation magnetization J_{sat}, the saturation remanence J_{rs} and the coercivity of remanence B_{cr}. The first two parameters are measured when the field is turned on; the last two are remanence measurements measured when the field is off. The domain state, and hence the magnetic particle size, can be roughly determined from ratios of these parameters.

In a Day plot (Day *et al.* 1977), J_{rs}/J_{sat} is plotted as a function of B_{cr}/B_c. Single-domain grains are considered to have J_{rs}/J_{sat} ratios greater than 0.5 and B_{cr}/B_c ratios less than 4, while multi-domain grains have J_{rs}/J_{sat} ratios less than 0.05 and B_{cr}/B_c ratios greater than 4. In between are grains that are said to be pseudo-single domain (PSD), multi-domain grains with just a few domains that behave magnetically like single-domain grains. They are typically around 1 micron in size (more or less). The Day plot is somewhat controversial to interpret, mainly because most samples fall in the PSD grain-size range. It became clear with more study that mixtures of different magnetic grain sizes affect the interpretation of the hysteresis parameter ratios. Another concern to keep in mind is that the Day plot was originally developed for titano-magnetites, and should only be used for magnetite and titano-magnetite magnetic mineralogies. Tauxe *et al.* (2002) have suggested that, instead of the Day plot to present the hysteresis data, J_{rs}/J_{sat} (sometimes called 'squareness' because it determines the 'fatness' of the hysteresis loop) is plotted as a function of the coercivity B_c or coercivity of remanence B_{cr} separately.

First-order reversal curve (FORC) diagrams, derived from detailed measurements of hysteresis loops, are the logical next step in measuring the hysteresis properties of a sample. These time-intensive plots result from the measurement of many hysteresis loops for a sample over a range of field strengths. FORC diagrams indicate the coercivities of the magnetic grains in a sample (i.e. the magnetic grain sizes) and the degree of magnetic interaction between the magnetic particles. See Tauxe (2010) for an excellent coverage of hysteresis loops, FORC diagrams and the use of the hysteresis parameters in rock and environmental magnetic studies.

Magnetic mineralogy

After magnetic mineral concentration and magnetic mineral grain size, the third main property measured in environmental magnetism studies is the magnetic mineralogy of a sample. Different magnetic minerals can be identified, usually without heating, by their different coercivities. The variation in the relative amounts of different magnetic minerals can be the result of changes in the depositional environment or source area or environmental changes in the source area. The relative amounts of these two magnetic minerals can be detected magnetically.

One of the best examples of using magnetic parameters to detect variations in the magnetic mineralogy of a sediment is the S-ratio. This ratio detects the relative amounts of a low-coercivity ferromagnetic mineral, i.e. magnetite, compared to the amounts of a high-coercivity antiferromagnetic mineral, i.e. hematite, in a sample. The sample is typically magnetically saturated, i.e. the intensity of IRM no longer increases as the sample is exposed to higher and higher DC magnetic fields in an IRM acquisition experiment, and then the sample is exposed to a field in the opposite direction called the backfield. Typically, a 0.3 T field is chosen for the backfield since this strength is the theoretical maximum coercivity of magnetite. The S-ratio is calculated as:

$$S\text{-ratio} = -\frac{IRM_{-0.3T}}{SIRM}$$

where SIRM is the saturation isothermal remanent magnetization. In the S-ratio experiment, the SIRM activates all the magnetic minerals (both the magnetite and the hematite) in a sample. The backfield magnetizes the magnetite in the reverse direction, leaving any hematite in the sample magnetized in the original direction. The ratio of these two IRMs indicates the relative proportion of magnetite to hematite. The negative sign for the backfield IRM in the equation above means that a sample containing only the ferrimagnetic phase (magnetite) will have an S-ratio of +1. A sample with only the antiferromagnetic mineral (hematite) will have an S-ratio of −1.

Variations in the S-ratio could be used to detect variations in the provenance of the sediments being studied. For instance, more antiferromagnetic minerals in lake sediments could indicate that the subsoil in a catchment is contributing more to erosion into a lake (Cioppa & Kodama 2003a). One point to keep in mind when interpreting the S-ratio is that the spontaneous magnetization of magnetite is about 200 times that of hematite, so small deviations from an S-ratio of 1 could mean respectable amounts of hematite in a sample. The S-ratio can also be used with different backfield strengths, designed to detect different magnetic mineralogies or magnetic grain sizes.

Even though heating is avoided in environmental magnetic studies because of the potential for causing chemical changes to the magnetic minerals in a sample, thermal behavior is a powerful tool for identifying magnetic mineralogy. Magnetite loses its magnetization at c. 580°C, hematite at c. 680°C and Fe sulfides lose their magnetization at c. 300°C. At these temperatures, the sub-atomic magnetic order in a crystal is overcome by the disordering influence of thermal energy and due to the expansion of the crystal lattice. The temperature at which the mineral loses its magnetization is referred to as the Curie temperature for magnetite and the Neel temperature for hematite, because hematite is an antiferromagnetic mineral. Goethite, another antiferromagnetic mineral important in environmental magnetic studies, has a Neel temperature of c. 120°C.

The so-called Lowrie test (Lowrie 1990) is often used to identify the magnetic mineralogy in a sediment. Samples are given orthogonal IRMs in different fields, usually a very high (c. 1 T), an intermediate-strength (c. 0.5 T) and a low-strength (c. 0.1 T) field. The sample is then thermally demagnetized. The test exploits both the thermomagnetic behavior and coercivities of different magnetic minerals to aid magnetic mineral identification. If the low field IRM (0.1 T) disappears at about 580°C this strongly suggests the presence of magnetite. High coercivity IRMs (1 T) that are gone at 680°C indicate hematite; if the IRM decreases at 120°C then goethite is suggested. Intermediate or low coercivities that are removed by 300°C temperatures indicate the presence the iron sulfides. Maghemite can be mistaken for a sulfide since it inverts to hematite at about 350°C, greatly reducing the intensity of magnetization. Discriminating maghemite from Fe sulfides is aided by re-exposing the sample to the pre-heating IRM fields. If much lower IRM intensities are acquired, maghemite is implicated since it has changed to hematite with its low spontaneous magnetization during heating. An increase in IRM strength, greater than before the heating, could implicate Fe sulfides since they usually oxidize to magnetite during the heating.

Magnetic parameter ratios

There are many different ratios of magnetic parameters used to detect environmental processes. We can only provide some of the most commonly used parameters and ratios here. The reader is directed to the environmental magnetism or paleomagnetism books cited above for more examples. Usually, the ratios or parameters employed will depend on the environmental process that is being investigated. For example, the goethite to goethite + hematite ratio has been suggested as a measure of the integrated climate, i.e. precipitation levels, in a sediment's source area (Harris & Mix 2002). Hematite and goethite in a sample can be detected by a series of magnetic measurements. Application of a very high field IRM, for instance 5 T applied in an impulse magnetizer, will activate all the magnetite, hematite and goethite in a sample. Alternating field demagnetization of the sample at 100 mT would then remove the magnetic contribution of most (if not all) of the magnetite in a sample, leaving only the hematite and goethite. Finally, either thermal demagnetization at a very low temperature (c. 130°C) would remove the goethite, leaving only the hematite, or a backfield of about 1 T would show the relative abundance of hematite to goethite as in the S-ratio for magnetite and hematite mentioned above. An alternating field demagnetization of 100 mT would then be needed again to remove the contribution of the magnetite reactivated by the 1 T backfield.

These measurements are easy enough to make for a large suite of samples, so the variation of this environmental process can be observed through time in a sedimentary section. The Neoprotereozoic Rainstorm Member of the Johnnie Formation displays a beautiful 5 m wavelength cycle of the magnetically determined goethite to hematite ratio interpreted by Kodama & Hillhouse (2011) to be about 100 kyr in duration, thus giving a high-resolution estimate of the duration of the Rainstorm member's deposition (Fig. 8.1).

DETECTING ENVIRONMENTAL PROCESSES

Some examples of how different environmental processes can be detected and studied using environmental magnetic measurements now follow. All environmental magnetic studies rely on the knowledge of

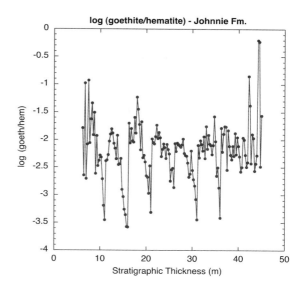

Fig. 8.1 Magnetically determined goethite to hematite ratio for the Rainstorm Member of the Johnnie Formation from Death Valley, CA showing a beautiful 5 m cycle. Kodama & Hillhouse (2011) interpret this cycle to be 125,000 years in duration (short eccentricity). (See Colour Plate 13)

basic rock magnetic principles and the ability to apply them creatively to different environmental conditions.

Lake sediment environmental magnetism

Many environmental magnetic studies are conducted on lake sediments because they provide excellent, nearly continuous records of environmental processes on the continents. They are an important complement to environmental magnetic studies of marine sediments since environmental processes on land can be compared and understood in the context of changes in the global ocean. Li et al. (2006, 2007) studied the environmental magnetism of Holocene-age sediments from White Lake in northern New Jersey and were able to detect changes in lake level through time. The changes in lake level are, of course, a record of long-term precipitation/evapo-transpiration changes in the area. The magnetic properties of two cores collected from the lake showed periodic high magnetic intensity layers separated by c. 1500 years of sediment deposition, based on a ^{14}C age model for the lake sediment

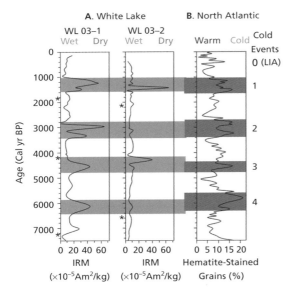

Fig. 8.2 Comparison of IRM peaks in White Lake sediment compared to cold events recorded in the North Atlantic by Bond *et al.* (2001). Figure from Li *et al.* (2007). Asterisks are the locations of ^{14}C samples used for dating the White Lake sediments. The IRM peaks are caused by oxidation of lake sediments during low stands of the lake water level. Y-X Li, Z Yu and KP Kodama, Sensitive moisture response to Holocene millenial-scale climate variations in the Mid-Atlantic region, USA, *The Holocene*, 17, 1, 3–8, 2007, Sage Publishing. (See Colour Plate 14)

(Fig. 8.2). The high-intensity layers were interpreted to be due to oxidation of the magnetic minerals in the lake during the exposure of lake sediments at times of low lake level. The increase in magnetic intensity due to oxidation was supported by observing sediment oxidation in the laboratory with sediment collected from deeper parts of the lake. The periods of low lake level observed magnetically coincided with cold periods in the North Atlantic based on North Atlantic sediment records of paleoclimate.

Cioppa & Kodama's (2003a) study of the Holocene sediments of Lake Waynewood in the Pocono Mountains of northeast Pennsylvania used the abrupt change in environmental magnetic parameters (ARM, SIRM, χ and their ratios, as well as the S-ratio) at 2900 years BP to indicate a significant change in watershed dynamics (Fig. 8.3). The change in parameters was

interpreted to indicate the start of the influx of material eroded from the watershed. This interpretation was bolstered by the appearance of anisotropy of anhysteric remanence (AAR) fabric at this time, indicating the initiation of the major inflow stream into the lake. These changes probably reflect a major change in the hydrologic regime in the area. The clear-cutting of the watershed at *c.* 100–200 years BP was also detected magnetically by large increases in the magnetic mineral concentration of magnetite, probably indicating the erosion of topsoil into the lake.

Kodama *et al.* (1997b) also looked at magnetic properties of recent lake sediments from several glacial lakes in the Pocono Mountains of northeastern Pennsylvania, including Lake Lacawac and Lake Giles. They observed an abrupt increase in ARM intensity in the top 10 cm of the sediment column that started at *c.* 1900 AD. SEM observation of magnetic extracts from the sediments indicates that framboidal magnetite microspherules are the culprit, and an indicator of the initiation of significant amounts of fossil fuel combustion in the northeastern United States for electrical power generation.

Oldfield (1990) had made similar observations in Big Moose Lake in the Adirondak Mountains of New York. The presence of antiferromagnetic minerals in the top 10 cm of one lake, also seen in soil samples from the lake's watershed, indicated erosion of the catchment due to European settlement in the area. Finally, time series analysis of magnetic mineral concentration variations identified a 50 year period in magnetic intensity variation similar to the same frequency of variation in historic rainfall data from the northeastern US over the past 150–250 years (Fig. 8.4). The implication is that magnetic mineral concentration variations could be used as a proxy for rainfall variations in a lake's watershed.

The last of our examples of how magnetic mineral parameters can aid paleoenvironmental studies of lake sediments comes from Hurleg Lake in China (Zhao *et al.* 2010). Zhao *et al.* used a multi-proxy record collected from the lake sediments to look at millennial-scale changes in lake level on the Tibetan Plateau to monitor fluctuations in the strength of the monsoon. Spikes in magnetic mineral concentration detected by ARM, SIRM, and χ and low ARM/SIRM ratios suggest more coarse-grained magnetic minerals entering the lake due to lower lake levels. This interpretation was made in concert with other paleoclimatic indicators, i.e. percentage carbonate and Mg/Ca ratios, and shows

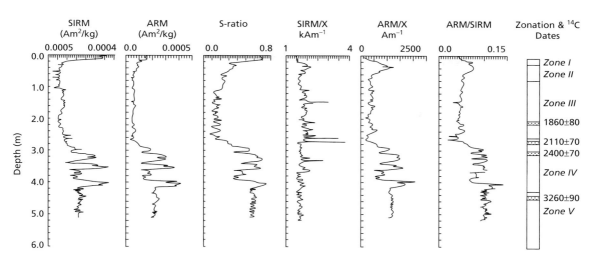

Fig. 8.3 Dramatic change in magnetic mineral parameters at a depth of *c.* 4.4 m (2900 years BP) in Lake Waynewood sediments. The large-amplitude variations in magnetic parameter then ends abruptly at a depth of *c.* 3 m (Cioppa & Kodama 2003a). MT Cioppa and KP Kodama, Environmental magnetic and magnetic fabric studies in Lake Waynewood, northeastern Pennsylvania, USA: Evidence for changes in watershed dynamics, *Journal of Paleolimnology*, 29, 61–78, 2003, Springer.

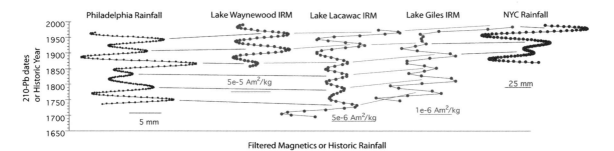

Fig. 8.4 Data from Kodama *et al.* (1997b) showing correlation between historic rainfall and IRM intensity of lake sediments from three Pocono Lakes (Waynewood, Lacawac and Giles). Data were filtered to remove high-frequency variations. KP Kodama, JC Lyons, PA Siver and A-M Lott, A mineral magnetic and scaled-chrysophyte paleolimnological study of two northeastern Pennsylvania lakes: records of fly ash deposition, land-use change, and paleorainfall variation, *Journal of Paleolimnology*, 17, 173–189, 1997, Springer. (See Colour Plate 15)

how environmental magnetic parameters can aid paleoclimatic studies of lake sediments.

Eolian and fluvial magnetic mineral sources

In some studies magnetic mineral parameters are used to detect eolian magnetic minerals. Oldfield *et al.*

(1985) found $\chi_{ARM}/SIRM$ ratios between 0.22 and 0.00102 m/A for the magnetite from eolian dust collected shipboard in the northern Atlantic Ocean and on the island of Barbados during the summer and winter seasons. Hounslow & Maher (1999) found very similar $\chi_{ARM}/SIRM$ ratios (0.25–0.00124 m/A) for the youngest marine sediments in cores from the Indian Ocean. χ_{ARM} is the ARM divided by the DC magnetic field used for application of the ARM. Hounslow and

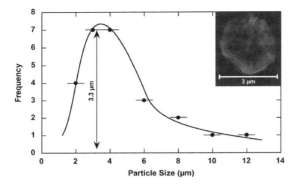

Fig. 8.5 Grain-size distribution of magnetic particles in the Cupido Formation and SEM image of a magnetic extract from the Cupido showing grain size and morphology of magnetic grains, suggesting an eolian source (Latta 2005).

Maher interpreted these data to indicate that the magnetite in these sediments was derived from eolian dust sourced from either the Arabian Peninsula or Somalia. Latta et al. (2006) took this approach further back in time by measuring the concentration of magnetite in Cretaceous platform carbonate rocks from Mexico, observing Milankovitch cyclicities in the ARM variations. They interpreted these variations as representing changes in global aridity in the Cretaceous, assuming that the magnetite was derived from eolian dust. This interpretation was bolstered by SEM examination of magnetic extracts showing magnetite grain sizes consistent with eolian dust (Latta et al. 2006; Fig. 8.5).

Similar χ_{ARM}/SIRM ratios were found for the Triassic platform carbonates of the Latemar in the Dolomites of northern Italy. The Latemar controversy will be covered in more detail later in this chapter, but the similarities to the Mexican work is that both units were deposited in platform carbonate environments with little detrital input from continental areas by runoff. Cyclicities in ARM intensity were observed in the Latemar rocks and based on their magnetic grain size and magnetic mineralogy (magnetite), and were interpreted to be derived from eolian dust. The assumption that little river-derived sediment is contributing to these platform carbonates can and should be checked with magnetic mineral parameters. There is little evidence in the literature for the magnetic grain size, and hence the χ_{ARM}/SIRM ratios of fluvially derived sediments. However, Zhang et al. (2008) studied estuarine sediments associated with the Yellow and Yangtze

Rivers in China and found χ_{ARM}/SIRM ratios of these fluvially derived sediments to be between 0.1 and 0.9×10^{-3} m/A. Kumar et al. (2005) conducted a rock magnetic study of Arabian Sea sediments of the coast of India, and found χ_{ARM}/SIRM ratios from 1.0 to 2.0×10^{-3} m/A for early Holocene sediments considered to have been derived from predominately fluvial sources. It is then apparent that eolian-derived magnetic particles have χ_{ARM}/SIRM ratios larger than that of fluvially derived particles. Since higher ratios indicate a smaller magnetic particle size, this observation is consistent with the grain size sorting expected due to transport by wind compared to that by water. It also shows the power of environmental magnetic parameters to detect the source environment of sediments and sedimentary rocks.

Loess, an eolian sediment derived from glacially eroded material, has been shown to carry an important environmental magnetic signal (e.g. see Evans & Heller 2003). The loess plateau of central China is the most important deposit of these sediments globally. Glacial and interglacial periods are recorded by intervals of weak and strong susceptibilities, respectively. The record can be tied unequivocally to the oxygen isotope record of glacial and interglacial periods in marine sediments by a magnetostratigraphy also carried by the loess sediments. The differences in susceptibility between glacial and interglacial periods are interpreted due to growth of secondary magnetic minerals in the warm, humid interglacial periods, thus enhancing the magnetic susceptibility of the interglacial paleosols. In fact, there has been an attempt by several workers (e.g. Heller et al. 1993; Maher et al. 1994) to calibrate the magnetic intensity of the paleosols to measure paleoprecipitation; the more rain, the more secondary growth of magnetic minerals and the stronger the magnetic intensity. Calibration is performed using modern-day gradients of precipitation across the loess plateau of China.

Marine sediments are also an important repository of environmental and climate-driven magnetic signals. The ability of environmental magnetic parameters to detect eolian-derived and fluvially derived sources has already been mentioned. The variations of the concentration of these magnetic minerals, measured primarily by susceptibility but also laboratory-induced remanences, have been used to monitor variations in terrigenous influx and have shown Milankovitch cycles and glacial–interglacial cycles (Bloemendal et al. 1988, 1992).

Environmental processes and DRM accuracy

Understanding the effects of environmental conditions and changes on the magnetic minerals in a sediment has important implications for the robustness and accuracy of the paleomagnetic signal, the overriding focus of this book. The source and primacy of the magnetic minerals are an important control on the age of a sediment's paleomagnetism. For instance, if the magnetic minerals are biogenic, they are secondary but probably formed soon after deposition of the sediment. If the magnetic minerals are clearly fluvially derived or from eolian sources, they are primary depositional magnetic minerals. If astronomically driven global climate changes are recorded by variations in magnetic mineral concentration, these variations could leak into the relative paleointensity records in marine sediments. This could happen because variations in magnetic mineral concentration, usually detected by ARM or IRM, are used to normalize the relative paleointensity record in sediments, assumed to be a record of geomagnetic field intensity changes.

Some workers have done detailed work to check whether astronomical variations in the paleointensity normalizer have affected relative paleointensity records from marine sediments (e.g. Channell *et al.* 2004). Such care is needed in all relative paleointensity studies to ensure that geomagnetic field variations are all that is being observed. Another consideration for the accuracy of a sedimentary paleomagnetic record is whether changes in paleoproductivity caused variations in the amount of organics in a sediment, and therefore affected the degree of reduction diagenesis and dissolution of magnetic minerals. Tarduno (1994) observed this effect in Pacific Ocean sediments. Dissolution due to redox fronts are seen with a 100 kyr periodicity, but offset in time from the 100 kyr glacial–interglacial intervals by 30–40 cm due to the lag between deposition and reduction diagenetic dissolution in the sediment column.

A final example of how environmental changes could affect the paleomagnetic signal comes from Chapter 4 on inclination shallowing. It was shown that clay content could affect the degree of compaction-caused inclination shallowing and possibly the post-compaction intensity of the NRM. These changes in clay content, driven by environmental changes, could therefore show up as cyclicities in inclination or paleointensity not related to geomagnetic field variations.

CASE STUDY: FORMATION OF BIOGENIC MAGNETIC MINERALS AT LAKE ELY

There are many examples of biomineralization, but it is beyond the scope of this book to attempt a comprehensive review. For the purposes of this chapter we're interested in a small subset of organisms that produce minerals, magnetic minerals that are iron oxides and sulfides formed by bacteria in lake and marine sediments. These bacterially produced minerals have been the focus of much study in the last several decades since they were first described (Blakemore 1975). Recently, a multi-cellular prokaryote from a coastal lagoon in the Yellow Sea region has been reported to produce magnetic minerals (Zhou *et al.* 2011), so it's apparent that the production of biogenic magnetic minerals is not restricted to bacteria. Since the 1980s, magnetoreception and biomineralization have been reported for fish, pigeons, bees and other higher animals (Kirschvink *et al.* 1985).

There are two types of bacterially produced magnetic minerals: those due to biologically organized mineralization (BOM) processes and those due to bacterially induced mineralization (BIM) processes. BOM processes are also called bacterially controlled mineralization (BCM). In BOM or BCM bacteria, chains of single-domain magnetite or greigite particles are created to act as magneto-receptors. The chains of small magnetic particles allow these micro-aerophilic bacteria to sense the Earth's weak magnetic field and use it to swim up and down along the magnetic field lines. At intermediate latitudes where most of these bacteria are studied, the Earth's magnetic field lines are steeply inclined to the horizontal. This ability allows the bacteria to stay near the oxic–anoxic interface (OAI). The OAI is typically in the uppermost part of the sediment column, but can be in the water column of lakes with anoxic bottom waters. The focus of the case study presented here is the production, preservation and environmental magnetic detection of magnetosomes in the freshwater environment of Lake Ely.

BIM minerals are produced extra-cellularly when bacteria use Fe for respiration. While the magnetosome chains of SD grains generated by magnetotactic bacteria are very stable magnetically, the magnetite produced in the BIM process are much smaller and are not stable magnetically; they are superparamagnetic in grain size (<20–30 nm for magnetite). The magnetization of superparamagnetic grains is caused by the same quantum mechanical sub-atomic processes as for larger magnetically stable grains, but only holds its direction for short periods of time (<several minutes) and reacts to applied fields similar to paramagnetic materials. Superparamagnetic grains therefore contribute to the susceptibility of a substance. The susceptibility of superparamagnetic grains can be detected by applying alternating magnetic fields at different frequencies. Both Tauxe (2010) and Butler (1992) provide excellent and detailed accounts of superparamagnetism.

Pertinent to the theme of this book is the consideration of whether the magnetosomes generated by magnetotactic bacteria contribute to the paleomagnetic remanence of a sediment or a sedimentary rock. The magnetofossils created by magnetotactic bacteria are found in many environments: lake sediments, hemipelagic marine sediments, deep-sea sediments, ancient limestones as far back as the Precambrian and even in Chinese loess (Chang & Kirschvink 1989; Evans & Heller 2003). As early as the 1980s, workers recognized that magnetosomes could become oriented by post-depositional processes to carry a strong paleomagnetic signal (Petersen *et al.* 1986). In a recent study Kobayashi *et al.* (2006) found that magnetosome chains maintained their linearity even after the cell walls surrounding the chains had been disrupted. Through TEM observations they found an intracellular organic sheath that may be responsible for stabilizing the chain structure. If chains could remain intact, or at least partially intact, they could provide strong, stable NRMs in marine and lake sediments. In fact, Roberts *et al.* (2011) recently provided evidence of bacterial magnetite preserved in Eocene pelagic carbonates from the southern Kerguelen Plateau, dominating their paleomagnetic signal. The chemical conditions were apparently just right in these organic-poor carbonate-rich pelagic sediments for magnetotactic bacteria to be present, but not dissolved after deposition by sulfate reduction diagenesis.

The importance of magnetosomes, produced in recent lake sediments, to the environmental magnetic record can be shown in detail from an environmental magnetic study of Lake Ely. In this study environmental magnetic parameters can be used to characterize the magnetic properties of the lake sediments and learn something about past environmental conditions in the lake and its watershed. The closest similar study is that of Salt Pond in Massachusetts (Moskowitz *et al.* 2008) but Salt Pond, as is obvious from its name, has

very saline waters. By contrast, Lake Ely is a typical freshwater lake from glaciated terrain in north-eastern Pennsylvania (41.760°N, 75.835°W, 380 m elevation; Fig. 8.6). The lake is relatively deep (24 m) surrounded by steep topography and has a small watershed. The

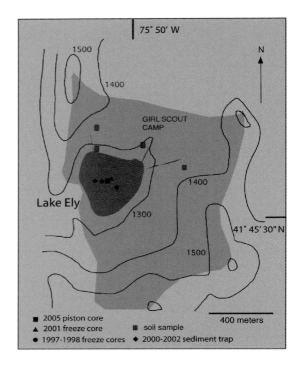

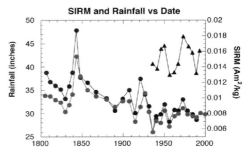

Fig. 8.6 Location map for Lake Ely in northeastern Pennsylvania and correlation between IRM intensity of recent lake sediments and local historical rainfall record. Red and blue circles indicate SIRM intensity for lake sediment samples collected at two successive years. Triangles show local historic rainfall measured near the lake. (See Colour Plate 16)

lake was initially targeted for environmental magnetic study because its sediment is rhythmically layered. ^{210}Pb dating indicates that these layers are annual varves and they allow high-resolution dating of the sediments, important for correlations with historic environmental records. The well-preserved laminations indicate that the lake sediments are not disturbed by organisms and therefore that the water near the lake bottom is anoxic throughout the year.

Kodama *et al.* (1997a, 1998) initially made the observation that variations in magnetic mineral concentration, based on SIRM intensity variations, correlate with historic variations in rainfall recorded nearby at Montrose, Pennsylvania over the past 70 years (Fig. 8.6). Careful measurement of the organic-rich dark layers of the varve couplets indicates that they have higher magnetic mineral concentrations than the light-colored silt layers in the varves (Kim *et al.* 2005). Furthermore, the silt layers contain more high-coercivity antiferromagnetic minerals from the watershed soils. While the correlation between magnetic mineral concentration and historic rainfall could indicate that the magnetic minerals in Lake Ely's sediments were derived from erosion of the catchment, environmental magnetic parameters suggest instead that the magnetic minerals were not formed in the watershed but must have been formed in the lake. In fact, the lake sediment magnetic minerals turned out to be dominated by magnetosomes created by magnetotactic bacteria living in the water column of the lake.

The dominance of magnetosomes in the lake's environmental magnetic record was shown by a combination of water column measurements. These measurements show that the OAI is situated at a depth of 15 m with a marked decrease in dissolved oxygen at that level. Furthermore, dissolved sulfide and Fe(II) increase below the sharp decrease in dissolved oxygen. At the OAI, the ARM of material filtered from the water suddenly increases, indicating the presence of stable, probably single-domain, magnetic mineral particles (Fig. 8.7). This of course suggests that magnetotactic bacteria are living at the OAI, confirmed from the TEM observation of magnetosome chains in the material filtered from the water column at this depth and lower in the water column (Fig. 8.7).

Bazylinski (2007, personal communication) has identified familiar species of magnetotactic bacteria in the water column as well as a new species of the genus *Magnetospirillum*. A sediment trap was installed in the deepest part of the lake, below the OAI, to see how and if the magnetosome chains produced by the magnetotactic bacteria in the water column were being transferred to the lake's sediments. TEM observations and magnetic mineral parameters measured from sediment collected in the trap, as well as from lake sediment sampled by a gravity core, show the presence of magnetosomes. This indicates that the dominant magnetic minerals of the lake sediment are indeed formed by magnetotactic bacteria in the lake. Furthermore, the initial observation of a correlation between local rainfall variations and magnetic mineral concentration would strongly suggest that the magnetotactic bacteria are responding to and recording rainfall variations in the area. If this mechanism is borne out it could provide a powerful paleoclimate proxy using environmental magnetic measurements of lake sediments, provided, of course, that the magnetic mineralogy of the lake sediment is dominated by magnetosomes.

With the confirmation that magnetotactic bacteria contribute the bulk of the magnetic minerals in the lake sediments, several techniques designed to detect magnetosome chains could be tested. Since the most definitive way of detecting magnetosomes involves time-consuming and somewhat difficult extraction of magnetic minerals, and then preparation and examination of the extract under a TEM, quicker and nondestructive magnetic tests would be highly preferable. Oldfield, a pioneer in the environmental magnetism field, has proposed a bivariate plot as a test for magnetosomes, χ_{ARM}/χ_{fd} versus χ_{ARM}/χ (Oldfield 1994) where χ_{fd} is the frequency dependence of susceptibility which, as mentioned before, can be used to detect superparamagnetic grains. An alternating magnetic field is used to measure susceptibility in most modern instruments so the samples don't pick up a viscous magnetization from the application of a DC magnetic field. The Lake Ely data was used in a modification of Oldfield's initially proposed bivariate plot with $\chi_{ARM}/$SIRM plotted versus χ_{ARM}/χ. The modification, according to Oldfield, is allowed for magnetic grains smaller than 0.07 μm, as would be expected for the SD grains that make up magnetosome chains. The modified Oldfield plot was used by Snowball *et al.* (2002) with magnetic data from Swedish lake sediments inferred to carry magnetosomes. The Lake Ely data fall perfectly in the same field as Snowball *et al.*'s data (Kim *et al.* 2005).

Moskowitz *et al.* (1993) proposed a diagnostic test for magnetosomes that is based on the ratio of a sample's magnetization when it is cooled to very low temperatures (20 K) in either a strong magnetic field (2.5 T) or

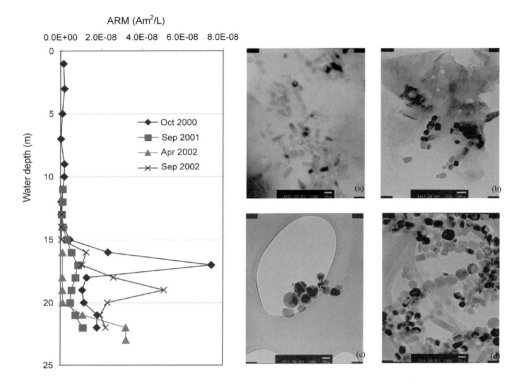

Fig. 8.7 ARM of material filtered from Lake Ely's water column at the oxic–anoxic interface (left) and TEM images of the filtered material showing magnetosomes created by magnetotactic bacteria. TEM images from Kim *et al.* (2005). B-Y Kim, KP Kodama and RE Moeller, Bacterial magnetite produced in water column dominates lake sediment mineral magnetism: Lake Ely, USA, *Geophysical Journal International*, 163, 26–37, 2005, Blackwell Publishing. (See Colour Plate 17)

a zero magnetic field. Magnetite goes through a transition in crystal structure at about 110 K that affects its magnetism. The magnetization of a sample decreases sharply when it is warmed up through the 110 K magnetic transition, called the Verwey transition. If it has been cooled in a strong magnetic field it drops more than if it was cooled in a zero strength field. The ratio of magnetization decrease is 2 or greater for magnetosome chains from samples of magnetotactic bacteria cultured in the laboratory. The test was much less successful for the Lake Ely sediments (Kim *et al.* 2005). The dark organic-rich layers from the lake sediment had weak Verwey transitions at 110 K while the light-colored silt layers saw virtually no Verwey transitions. The Moskowitz *et al.* (1993) ratios of field-cooled to zero-field-cooled magnetization decreases were only about 1.5, suggesting that magnetosome chains were

not present in the samples despite TEM observations of the chains.

Unmixing of the low-temperature magnetic behavior using numerical modeling (Carter-Stiglitz *et al.* 2001) however suggests that there are about twice as many magnetosome chains in the dark organic-rich layers as in the silt layers (*c.* 60% compared to *c.* 30%). The failure of the test is primarily due to the additional magnetic minerals, besides magnetite, in the lake sediments (antiferromagnetics from the watershed and greigite produced biogenically).

Moskowitz *et al.* (1993) also produced a diagnostic test for magnetosomes that could be conducted at room temperature, thus not requiring the specialized equipment needed to cool samples to nearly absolute zero and simultaneously measure their magnetization while a strong field is applied. This test involves the plot

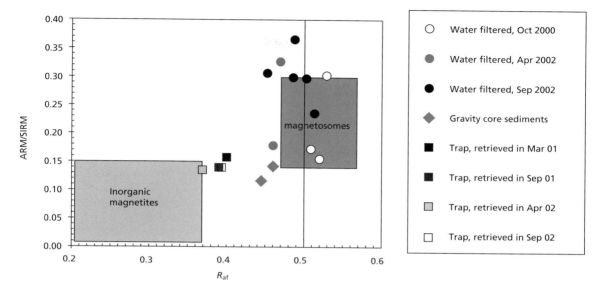

Fig. 8.8 ARM/SIRM versus R_{af} for Lake Ely sediments, sediment trap material and material filtered from the water column. All the results show that this measurement is a good test for detecting magnetosomes made by magnetotactic bacteria. Figure from Kim *et al.* (2005). B-Y Kim, KP Kodama and RE Moeller, Bacterial magnetite produced in water column dominates lake sediment mineral magnetism: Lake Ely, USA, *Geophysical Journal International*, 163, 26–37, 2005, Blackwell Publishing.

of the ARM/SIRM ratio of a sample versus a quantity denoted R_{af}, the crossover between a sample's IRM acquisition curve and the alternating field demagnetization of its SIRM. It was initially proposed by Cisowski (1981) as a test for magnetic interactions between particles; values less than 0.5 indicated magnetic interactions. However, the magnetosome chains act like big non-interacting single-domain particles and R_{af} values greater than 0.5 are diagnostic of magnetosome chains, as are ARM/SIRM ratios greater than *c.* 0.15. Samples containing magnetosome chains therefore plot in the upper right-hand part of the ARM/SIRM versus R_{af} plot (Fig. 8.8) and detrital magnetite particles in the lower left. This test worked beautifully for the Lake Ely samples, showing magnetosome chains for the material filtered from the water column and the lake sediment collected by the gravity cores. It is interesting that the material caught in the sediment trap plots between the magnetosome chain region and the region for inorganic magnetite particles, based on the data of Moskowitz *et al.* (1993).

All the work discussed up to this point was based on measurements of material filtered from the water column, collected in a sediment trap at the bottom of

the lake and in gravity cores from the top 30 cm of the sediment column. Since it was demonstrated that magnetosomes dominated the magnetism of the lake sediments and that the concentration of magnetosomes appeared to record historic variations in rainfall, it was important to determine whether the magnetosome concentration variations would be preserved in the sedimentary record over longer time periods than just 100–200 years.

To answer that question, Robert Moeller collected a 125 cm long Livingstone piston core through the ice in the winter of 2005. Standard environmental magnetic measurements were made on detailed sampling of the core. SIRM and χ were measured every centimeter and ARM and the *S*-ratio every 3 cm. In addition to these magnetic measurements, the concentration of organic material was quantified by loss-on-ignition (LOI) measurements downcore. The main finding of this work was that the magnetic concentration, presumably of the magnetosomes in the lake's sediment, decreased significantly to a depth of 75 cm in the core (Fig. 8.9). *S*-ratio measurements show a greater contribution from high-coercivity antiferromagnetic minerals as the overall magnetic mineral concentration decreases

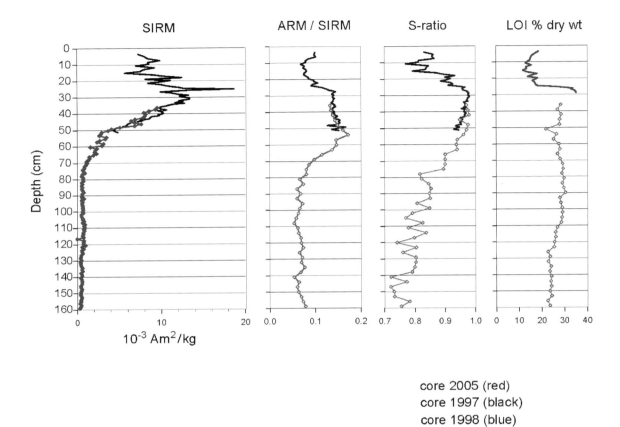

Fig. 8.9 Magnetic parameters measured for Lake Ely sediment collected by piston core (shown in red) compared to sediment collected from gravity cores (black). The parameters show evidence of reduction diagenesis causing dissolution of the magnetosomes generated in the water column. (See Colour Plate 18)

downcore, suggesting that the magnetic minerals from the lake's watershed soils are more important in the lake's sediment. The ARM/SIRM ratio also decreases over this depth range showing that the magnetic grain size increases as the concentration decreases.

The evidence is strong that reduction diagenesis, due to the high concentration of organic material (c. 25%), has dissolved away most of the magnetosomes that were deposited in the lake. This conclusion still needs to be tested by detailed TEM examination of magnetic mineral extracts above and below the precipitous decrease in magnetic intensity downcore. It is the most likely explanation given the high organic content of the sediment, but there is still the possibility that magnetosome production 'turned on' at a depth of 75 cm.

Varve counting suggests that dissolution occurs by about 1200 years after deposition, in sediment that was deposited at about 800 CE. The settlement of the lake watershed at 1820 CE based on historical records is marked by a sharp decrease in organic matter (decrease in LOI) due to increased erosion from the watershed and a commensurate increase in magnetic mineral concentration. It is this increase in material eroded from the watershed that increases the thickness of the varves and allows detailed magnetic examination of the dark organic-rich layers of the varves in comparison to the lighter silt-rich layers.

The environmental magnetic properties of the lake sediments were characterized in more detail, both at shallow depths before dissolution had occurred and at depths greater than the large loss in magnetic intensity. Kruiver *et al.* (2001) have provided an important method of modeling the first derivative of the IRM acquisition curves that is a standard way for identifying the magnetic mineralogy of a sample based on its different coercivity components. In the Lake Ely samples, IRM acquisition modeling shows two distinct coercivity components at low coercivities less than 100 mT (Figs 8.10 and 8.11). These are very similar to Egli's (2004) observation of two different

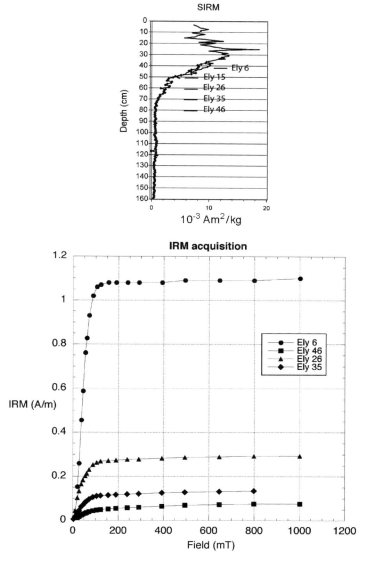

Fig. 8.10 IRM acquisition curves for Lake Ely sediment collected at different depths in the 2005 piston core. The top figure shows the location, with respect to the decrease in intensity downcore, of the samples used for the IRM acquisition experiments shown in the bottom figure.

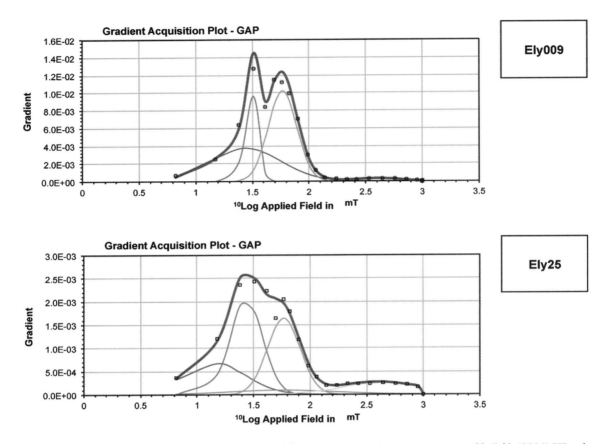

Fig. 8.11 IRM acquisition modeling (Kruiver *et al.* 2001) showing two coercivity components, possibly Egli's (2004) BH and BS magnetosomes in Lake Ely sediments. (See Colour Plate 19)

magnetosome coercivity components which he detected in other lake sediments and refers to as BH (biogenic hard) and BS (biogenic soft). However, the resolution of the IRM acquisition curves in the coercivity range <100 mT is poor. Taking the first derivative makes the data even noisier. The primary reason for this is that most laboratories use an impulse magnetizer, that discharges a capacitor through a coil surrounding the sample, to magnetize a sample for IRM acquisition experiments. The set-up is perfect for higher fields, but harder to control the field strength for magnetic fields <100 mT. The Lake Ely samples were therefore given partial ARMs with the DC field applied in increas-

ingly higher-strength alternating fields. This ARM acquisition experiment allows better resolution of the coercivity components in the <100 mT field range. Modeling of the ARM results shows that two distinct coercivity components could be distinguished in the <100 mT field range for samples collected at depths of 35 and 45 cm in the piston core. These most likely represent the BH and BS magnetosome components of Egli (2004) and could suggest contributions from magnetosomes with different morphologies. It is interesting that only one coercivity component dominates the ARM acquisition modeling at the shallowest depths in the lake sediments, suggesting that reduction diagenesis dissolution allows the observation of the BS and BH components at greater depths. This observation also suggests that different magnetosome morphologies are resistant to diagenesis by different degrees.

Chen *et al.* (2007) studied magnetosomes from laboratory cultures and from natural lake sediment with FORC diagrams to better determine the signature of magnetotactic bacteria. Subsequently, Egli *et al.* (2010) studied a sample from shallow in the sediment column of Lake Ely (6 cm depth in the piston core). The resulting FORC diagram shows no evidence of interactions between the particles. The FORC diagram shows a very tight distribution around the H_c-axis (where H_c is coercivity) and the magnetosome chains, demonstrated to be present from the earlier TEM observations of magnetic extracts, act as if they were non-interacting single-domain particles. The FORC diagram for a sample deeper in the sediment column (26 cm depth) shows a greater spread around the H_c-axis, indicating magnetic interaction between grains, possibly due to the disruption of magnetosome chains as reduction diagenesis proceeds.

Bob Kopp made ferromagnetic resonance (FMR) absorption spectra measurements on samples from the Lake Ely piston core. Since FMR measures the effective magnetic field in a sample, including magnetic particle anisotropy and magnetic interactions, it is perfect for detecting the presence of magnetosome chains produced by magnetotactic bacteria. Kopp *et al.* (2006) made FMR measurements of laboratory-grown magnetotactic bacteria, but measurements of natural lake sediment samples with demonstrated magnetosomes would be a good demonstration of the applicability of the technique. The FMR absorption spectra from samples collected at different depths from the Lake Ely piston core (Fig. 8.12) show the presence of magnetosomes, particularly in the top of the core. The amplitude of the spectra decreases downcore as reduction diagenesis affects the samples to a greater degree. The effective field (g_{eff}) has a value of *c.* 2.0, down slightly from that of pure magnetite (2.12); the anisotropy factor (*A* factor) degrades downcore. These spectra are consistent with the known presence of magnetosomes in the samples.

Kopp *et al.* (2007) demonstrated the presence of magnetosomes in the sediments recording the Paleocene–Eocene thermal maximum collected from a core drilled in coastal plain sediments from southern New Jersey. For this work they showed that in a plot of FMR parameters ΔB_{FWHM} (an empirical parameter measured from the FMR absorption factor, defined as the field *B* at full-width at half-maximum amplitude of the spectrum) versus *A*, magnetosome chains had lower values of these parameters and detrital magnetite had higher values. When the Lake Ely data are plotted on a ΔB_{FWHM} versus *A* plot (Fig. 8.13), the data fall close to the range of likely magnetofossils. Finally, modeling of the FMR spectra by Kopp shows that the spectra are consistent with the two different coercivity components for magnetosomes, BH and BS, originally observed by Egli (2004).

The environmental magnetic study of Lake Ely presented in detail here shows that environmental magnetic parameters can detect the presence of magnetosomes generated by magnetotactic bacteria, even the two different coercivity components (BH and BS) initially observed by Egli (2004). Modeling of IRM and ARM acquisition data is particularly sensitive to detection of magnetosomes. FMR absorption spectra are an entirely different way to detect magnetosome chains in natural lake sediment containing magnetosomes. Reduction diagenesis removes the magnetosome record almost completely by about 1200 years after deposition and suggests that reduction diagenesis can occur relatively soon after the deposition of lake sediments. The diagenesis at Lake Ely occurs within about 1 m of the sediment–water interface in the topmost part of the sediment column.

For Lake Ely, lake sediments dominated by magnetotactic bacteria magnetosomes could provide a record of paleorainfall variations, at least until the magnetosomes are completely destroyed by reduction diagenesis. In the topmost post-settlement sediments, the changes in the concentration of the magnetosomes follows the historic record of local rainfall. The ARM intensity of sediment trap material collected once a year in the fall over a three year period shows a good correlation with the rainfall that occurs after the ice leaves the lake in the spring, further bolstering the potential paleoclimate usefulness of magnetosome-dominated sediment (Fig. 8.14). The magnetic mineral concentration variations may possibly be caused by the amount of nutrients washed into the lake by different amounts of rainfall.

The environmental magnetic study of Lake Ely shows one way that magnetic mineral measurements can provide powerful indicators of past environmental

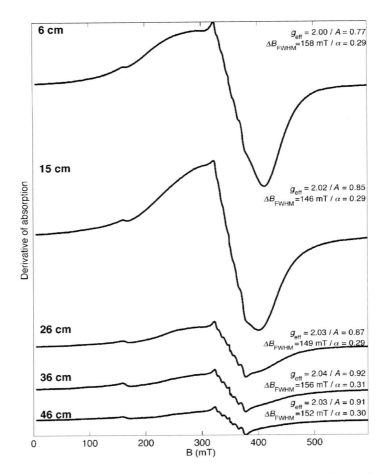

Fig. 8.12 FMR spectra for Lake Ely sediment collected at different depths in the piston core. The amplitude of the spectra decreases as the sediment's magnetic intensity decreases due to reduction diagenesis. Data collected by Bob Kopp.

changes in a lake and a lake's watershed. The standard measurements of magnetic mineral concentration, magnetic mineral grain size and magnetic mineral type can be used creatively, supplemented by non-magnetic observations, to reconstruct the local environmental past.

ROCK MAGNETIC CYCLOSTRATIGRAPHY

Rock magnetic cyclostratigraphy is a new application of environmental magnetic parameters to ancient sequences of sedimentary rocks. Typically, environmental magnetic parameters measuring the concentration, grain size and type of magnetic minerals are applied to contemporary or recent lake or marine sediments, soils or paleosols. In the simplest application of rock magnetic cyclostratigraphy, magnetic mineral concentration variations are used to detect orbitally driven global climate change at millennial or greater timescales, i.e. Milankovitch cycles, in order to provide a high-resolution chronostratigraphy to a sequence of sedimentary rocks. Milankovitch cycles are the subtle variations (c. 7%) in solar radiation input to the top of the Earth's atmosphere, i.e. insolation, at different lati-

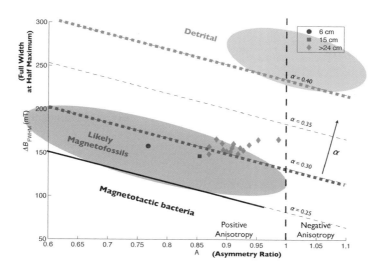

Fig. 8.13 Plot showing that FMR can identify magnetofossils. Data from Lake Ely is plotted with the green diamonds. Data and plot from Bob Kopp. (See Colour Plate 20)

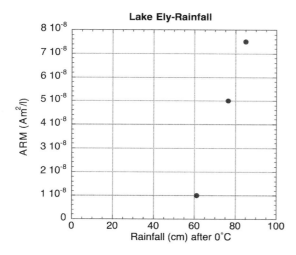

Fig. 8.14 Strong linear correlation between the ARM of material collected from Lake Ely's water column at the OAI over three years and the local rainfall after the daily temperature rose above 0°C and the lake was presumably no longer ice covered. These results suggest that the population of the magnetotactic bacteria varies as a function of rainfall washing material into the lake.

tudes. They are caused by perturbations in the Earth's orbital motions due to the gravitational pull of the Moon, the Sun and the planets. The theoretical effect of these perturbations on insolation can be calculated theoretically (Laskar *et al.* 2004, 2011) for the last 40–50 million years and, with reasonable accuracy, back 65 million years. Beyond this time, precise calculations are more problematic because of the chaotic evolution of planetary orbits.

There are many observations of rhythmic bedding in sedimentary rock sequences that, in the field of cyclostratigraphy, are assumed to be caused by Milankovitch cycles. The global climate change that is presumably driven by astronomically forced insolation changes is then recorded by lithologic, geochemical or isotopic variations in the rock record. The earliest study is that of Hays *et al.* (1976) in which oxygen isotope and sea-surface temperature fossil assemblage variations in Late Pleistocene marine sediments from the southern ocean can be correlated to Milankovitch periodicities. Hinnov (2000) gives a more recent review of the cyclostratigraphic literature in which she refers to studies of Pleistocene palynological series from lake sediments, loess, biogenic silica records from

Lake Baikal, ice cores from the polar regions and mid-latitude continental glaciers and lake water depth records from the Triassic of the Newark Basin in eastern North America (Kukla *et al.* 1990; Hooghiemstra *et al.* 1993; Mommersteeg *et al.* 1995; Olsen 1997; Thompson *et al.* 1997; Williams *et al.* 1997). Hinnov (2000) indicates nearly continuous records of orbitally forced stratigraphy back to the Oligocene.

The astronomically forced Milankovitch cycles which we attempt to detect magnetically in the rock have periods of 20 kyr, 40 kyr, 100 kyr and 400 kyr. These variations are due to changes in the direction of the Earth's spin axis in space (precession), the degree of tilt of the Earth's spin axis with respect to its orbital plane around the Sun (obliquity) and the ellipticity of its orbit around the Sun (eccentricity). Over the past million years, precession has had an average period of 21 kyr but is actually made up of three major periods (19 kyr, 22 kyr and 24 kyr) and is due to the gravitational pull on the Earth's tidal bulge that causes its spin axis to precess. Precession simply changes where the seasons fall in Earth's orbit around the Sun and acting alone could not, if the Earth's orbit were perfectly circular, change insolation. However, when it is coupled with ellipticity changes in the Earth's orbit, it has a marked effect on insolation. If for example northern hemisphere winter occurs at aphelion, when the Earth is furthest from the Sun in an elliptical orbit, a very cold winter results. On the other hand, if northern hemisphere winter occurs at perihelion, Earth's closest approach to the Sun, northern hemisphere winter is moderated. Therefore, precession only effects insolation by being modulated by eccentricity.

Eccentricity, the ellipticity of the Earth's orbit, has varied with a nominal 100 kyr period (short eccentricity) and a 405 kyr period (long eccentricity) over the past million years. Short eccentricity is really made up of two periods close to 125 and 95 kyr. By itself, eccentricity only causes very small (0.1%) variations in insolation; when coupled with precession the effects are much bigger (several percent).

Obliquity is the third significant orbital variation in insolation. It has had a period of 41 kyr over the past million years and is due to small variations in the tilt of the Earth's spin axis with respect to the plane of the ecliptic, the plane defined by the Earth's orbit around the Sun. The variation is small (from 21.5 to 24.5°) and the Earth's tilt is currently about 23.5°. The tilt of the Earth's spin axis is what causes the seasons.

The periods of these variations change back through geologic time. Eccentricity (particularly long eccentricity, 405 kyr) stays fairly constant but the periods of precession and obliquity become much shorter, mainly because the Earth's rotation rate was faster in the past and the Moon was closer to the Earth. As the Earth's rotation has slowed down and the Moon has moved further away, precession and obliquity periods have increased.

An astronomically forced interpretation of rhythmic lithologic variations and rock magnetic variations is not always straightforward and may not always be warranted. One of the greatest stumbling blocks to an orbitally forced interpretation of cyclostratigraphy is understanding or accepting how minute variations in insolation at the top of the atmosphere can cause essentially linear changes in global climate, given the complexity and non-linearity of Earth's climate system (Rial *et al.* 2004).

There are some that argue, quite strenuously, that orbitally forced stratigraphy does not exist and any patterns observed are due to circular reasoning or unwarranted assumptions (Bailey 2009). Bailey (2009) takes issue with the tuning of cyclostratigraphic records to a particular Milankovitch frequency, a standard cyclostratigraphic analytical technique (see for example Preto *et al.* 2001) as a blatant example of circular reasoning. We will demonstrate tuning and how it can be used to identify Milankovitch forcing (the Cupido and Arguis formations). The resulting evidence of orbitally forced stratigraphy can be convincingly strong, but there are also examples where tuning has somehow produced the appearance of Milankovitch cycles despite being in conflict with more compelling evidence that they do not exist (the Latemar controversy).

Another pitfall to the assumption of orbitally forced cyclostratigraphy is the potential presence of autocycles in a given depositional environment. Autocycles in a platform carbonate environment will be considered in our coverage of the rock magnetic cyclostratigraphy of the Cupido Formation (see section on Cupido Formation below). Finally, the complexities of sediment transport to the depositional basin may work to 'shred' any climate or environmental signal so that energy injected into a system at one frequency can be smeared across a range of frequencies. High frequencies are much more prone to being affected (Jerolmack & Paola 2010).

Given all these problems, it might be tempting to simply throw up hands in despair and avoid the whole topic. The best approach is to assume that patterns in Nature probably mean something, and should be studied. Since most systems in Nature are only incompletely understood, however, always remember that the observer could sometimes be fooled.

If we however accept that the evidence is reasonably strong that sedimentary facies respond in some way to changes in Earth's climate and that Milankovitch forcing of global climate occurs, it is not a great leap to think that environmental magnetic parameters will somehow record orbitally forced climate cycles. Because the timescale of the sedimentary record is best suited to study millennial-scale cyclicity, rock magnetic cyclostratigraphy should be perfect for the detection of Milankovitch-scale cyclicity.

In our treatment of inclination shallowing using magnetic anisotropy, we avoided the use of magnetic susceptibility because it has many sources (diamagnetic, paramagnetic and ferromagnetic minerals) and is therefore a complicated signal to untangle. We follow the same philosophy with rock magnetic cyclostratigraphy, although susceptibility records of Milankovitch cycles have been reported in the literature – particularly for deep-sea marine sediments – which have a fairly straightforward depositional history (Hays *et al.* 1976; Bloemendal & de Menocal 1989; Mayer & Appel 1999). Remanence, particularly the remanence of a specific magnetic mineral and a specific magnetic grain size, can be used to construct a simpler measure of magnetic particle concentration, the primary parameter used in rock magnetic cyclostratigraphic studies. ARM data series for low-coercivity depositional magnetite or IRM data series for high-coercivity minerals such as hematite can be used to detect orbitally forced climate variations.

Other environmental magnetic parameters can be used to detect more complicated signals. For instance, the S-ratio can be used to quantify the changes in relative amounts of magnetite and hematite, or the ARM/SIRM or ARM/χ ratios can detect changes in magnetic grain size. More complicated parameters can be constructed on a case-by-case basis to measure other climate-sensitive proxies, such as the goethite to hematite ratio (see Fig. 8.1). While the response of sedimentary facies to cyclic climate changes can be difficult to measure or interpret, for instance subtle changes in sediment grain size due to either runoff variations or small changes in relative sea level, magnetic measurements can be very sensitive to these subtle changes. Moreover, magnetic measurements are relatively quick to make when compared to non-magnetic measurements (e.g. chemical, isotopic) and non-destructive. One magnetic measurement is in effect counting thousands of magnetic particles in a sample, which is one of the reasons for the sensitivity of a magnetic measurement. The short measurement time for one sample (*c.* 1–2 minutes) allows the collection of hundreds–thousands of measurements from one stratigraphic section.

Often the cyclic signals are seen in the data series in rock magnetic measurements, but time series analysis is needed to quantify the periodicities and sophisticated noise analysis is used to determine the statistical significance of the frequency peaks in the power spectrum. In a typical rock magnetic cyclostratigraphic study, a relatively thick stratigraphic section (tens to hundreds of meters) is sampled at very close intervals (sub-meter scale) and a rock magnetic parameter (ARM, SIRM, IRM) is measured for the hundreds–thousands of samples collected. Ideally, some independent measure of time needs to be established; standard magnetostratigraphy is one of the best techniques to do this. An independent time measure is critical so that Milankovitch periodicities (20 kyr, 40 kyr, 100 kyr, 400 kyr) can be unequivocally identified. Ratios of periodicities (1 : 2 : 5 : 20, particularly the 5 : 1 eccentricity to precession ratio) can be used to tentatively identify Milankovitch cycles, but this is not always foolproof (see the section on 'Latemar controversy' below).

One of the difficulties of rock magnetic cyclostratigraphy is the determination of the rock magnetic sampling interval. Time series analysis theory dictates that the highest frequency detectable in a data (time) series has a period of twice the sampling interval. This is called the Nyquist frequency and means that, if you want to detect precession in the stratigraphic section, you need to sample at least once every 10,000 years. For good resolution of precession it would be better to sample every 5000 years. For this, of course, you need a rough idea of the sediment accumulation rate for the section being studied. This could come from a previous magnetostratigraphic study or from a detailed biochronology for the section. At the worst sediment accumulation rates can be estimated from depositional environments and lithologies, but a preliminary pilot study may be needed before a detailed rock magnetic cyclostratigraphic sampling is completed.

Time series analysis used in rock magnetic cyclostratigraphy uses mathematical techniques that are essentially modifications of Fourier analysis. These mathematical analyses break the data series down into its component frequencies and assume that the repeating cyclicities are in the form of sinusoids. It's best to provide a rough estimate of time to the data series being studied, so that when the data series is converted into the frequency domain it is a power spectrum that is a function of time frequencies (e.g. Weedon 2003; fig. 1.3). If this isn't possible, a magnetic parameter data series that is a function of stratigraphic position can be converted to a spatial frequency power spectrum for initial evaluation. The multi-taper method (MTM, Thomson 1982) is typically used for time series analysis and the significance of the spectral peaks is determined from their emergence above robust red noise calculated for the spectrum (Mann & Lees 1996). The basic techniques will be illustrated in the following case studies.

CUPIDO FORMATION: CRETACEOUS OF MEXICO

The first example of a rock magnetic cyclostratigraphic study is the work of Latta *et al.* (2006) on the lower Cretaceous Cupido Formation from north-eastern Mexico. The Cupido Formation is a 940 m thick sequence of marine platform carbonates deformed during the Laramide in the Coahuila marginal fold-thrust province. Latta *et al.* (2006) sampled the Cupido at two localities: Potrero Chico and Potrero Garcia. The Garcia rocks were deposited in an inner shelf environment and the Chico rocks in a middle shelf environment. At both localities, there are meter-scale sequences of facies that indicate repetitive upward shallowing of the depositional environment. This behavior has been observed in other platform carbonate stratigraphic sections, notably the Latemar cycles in the Triassic of northern Italy that we will discuss in the section on the Latemar controversy below.

One obvious interpretation of the upward shallowing facies is that relative sea level is bouncing up and down. The favored cyclostratigraphic interpretation is that an oscillating outside forcing function, i.e. orbitally forced climate change, is driving sea-level fluctuations at Milankovitch frequencies. An alter-

nate explanation is however possible: that autocycles are responsible for the repetitive upward shallowing sequences. For platform carbonates the autocycles could be caused by slow subsidence of the platform, perhaps due to cooling of the lithospheric plate, providing accommodation space for the biogenic production of carbonate. When water depth hits some critical depth, carbonate can be produced relatively quickly until the space is filled and thus relative sea-level shallows until supratidal environments are reached. This sequence repeats itself as the platform slowly 'drowns'. There is no need to appeal to orbitally forced climate change to create the repeating facies and there is no 'climate signal' to be found.

Rock magnetic cyclostratigraphy was used to study the Cupido shallowing upward sequences in more detail and from a different perspective. Time could not be easily assigned to the sequence by magnetostratigraphy since its Aptian age puts it at the beginning of the Cretaceous normal polarity superchron, when the geomagnetic field did not reverse for tens of millions of years. A sequence stratigraphy analysis of the rocks indicates its Aptian age (*c.* 120 Ma) and suggests that the formation was deposited over about 3.6 million years, indicating that the shallowing upward cycles have a non-Milankovitch duration of about 73 kyrs. The 'fourth-order' stratigraphic sequences in the section would have a 730–900 kyr duration, again not a Milankovitch periodicity. Furthermore, no 5 : 1 bundling of the sequences is observed.

The Cupido Formation was sampled at high resolution using a 10 cm sample stratigraphic spacing for ARM measurements. Many samples were taken within each of the shallowing upward sequences; 140 m were sampled. A beautiful rock magnetic cyclostratigraphy resulted (Fig. 8.15) showing both long and short cycles that could be correlated between the Chico and Garcia localities. Furthermore, the longer wavelength maxima seen in the ARM data are in phase with the sequence boundaries determined from sequence stratigraphic analysis. Another important observation is that the ARM cyclicities are decoupled from cycles defined by the lithologic facies (Fig. 8.16), indicating that they arise independently. Rock magnetic analysis indicates that the ARM is carried by magnetite and it has a grain size, based on both the ARM/SIRM ratio and examination of a magnetic separate by SEM (Fig. 8.5), consistent with eolian dust. The data suggest that the ARM is a record of dust being blown into the sampling locality during the Early Cretaceous, independent of the

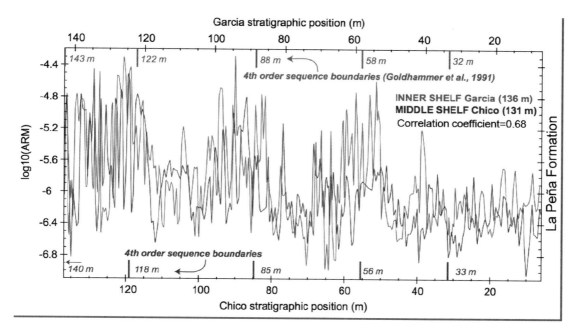

Fig. 8.15 ARM cyclostratigraphy for the Cretaceous Cupido Formation of northeastern Mexico. The ARM variations show long period and short period cycles. The fourth-order sequence boundaries identified by sequence stratigraphy match the maximum of the long-period variations in ARM. Figure from Hinnov *et al.* (2009). LA Hinnov, D Anastasio, D Latta, K Kodama and M Elrick, Milankovitch-controlled paleoclimate signal recorded by rock magnetics, Lower Cretaceous Platform carbonates of northern Mexico, American Association of Petroleum Geologists, Search and Discovery Article #40388, 2009, AAPG. (See Colour Plate 21)

shallowing upward cycles. Finally, if the sequence boundaries are assumed to indicate the 405 kyr long eccentricity cycle, thus assigning time to the ARM data series and essentially tuning the ARM data series to long eccentricity, an MTM spectral analysis shows spectral peaks at 100 kyr, 44 kyr and 20 kyr, expected Milankovitch frequencies (Fig. 8.17). The intepretation made by Latta *et al.* (2006) is that the ARM is a record of changing global aridity due to the influence of precession (20 kyr cycle) on the monsoon.

ARGUIS FORMATION

Another example of a rock magnetic cyclostratigraphic study comes from the study by Kodama *et al.* (2010) of the Eocene Arguis Formation in the Spanish Pyrenees. The near-shore marine marls of the Arguis were sampled at high resolution (0.75–1.5 m sampling

interval) for over 800 m of section to develop a high-resolution chronostratigraphy to study fold growth rates. A magnetostratigraphy was also measured for the Arguis so that absolute time would constrain the periodicities observed in the rock magnetic cyclostratigraphy. The magnetostratigraphy provides coarse time control and the cyclostratigraphy is designed to provide high-resolution (c. 20 kyr) time for the Arguis growth strata. The magnetostratigraphy was tied to the Geomagnetic Polarity Time scale (Gradstein *et al.* 2004) from a detailed biostratigraphy developed for the Arguis Formation. This is a very well-constrained study and is a good test of the reality of Milankovitch forcing of magnetic mineral parameters in a marine sedimentary section.

Rock magnetic measurements indicate that magnetite is the primary depositional magnetic mineral, but there are also Fe sulfides (most likely greigite) present in the rocks that are most certainly secondary and

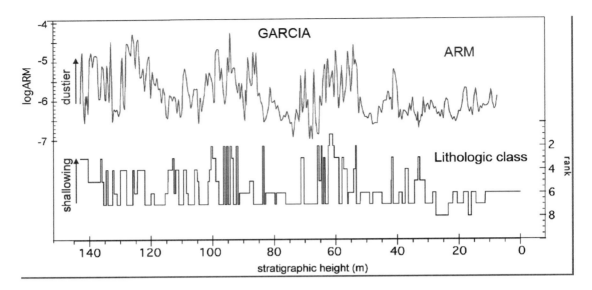

Fig. 8.16 ARM cyclostratigraphy from the Potrero Garcia locality of the Cupido Formation shows more structure and higher-frequency cycles than the depth rank series determined from the facies (Hinnov *et al.* 2009). The depth rank series is constructed by assigning a 'depth rank' for the different facies, i.e. quantifying the facies interpretation of water depth of deposition. For this study, lower depth ranks indicate shallower water depths. LA Hinnov, D Anastasio, D Latta, K Kodama and M Elrick, Milankovitch-controlled paleoclimate signal recorded by rock magnetics, Lower Cretaceous Platform carbonates of northern Mexico, American Association of Petroleum Geologists, Search and Discovery Article #40388, 2009, AAPG. (See Colour Plate 22)

formed by reduction diagenesis. The ARM cyclostratigraphy of the Arguis indicates that the secondary Fe sulfides do not swamp the depositional signal of the magnetite. When the ARM cyclostratigraphy is tied without tuning to time using the magnetostratigraphy, periodicities at Milankovitch frequencies emerge from the MTM spectral analysis of the ARM time series (Fig. 8.18). Strong peaks that rise above the 99% confidence limits of the robust red noise (Mann & Lees 1996) occur at low frequencies of *c.* 355 kyr, 232 kyr and 133 kyr. The 355 kyr and 133 kyr peaks are close to long (405 kyr) and short (125 kyr) eccentricity, respectively. A cluster of peaks near 5 m (24, 23, 21 kyr) suggests the encoding of precession by the ARM as well. Several other peaks also occur in the power spectrum, most notably at 89 kyr, 63.5 kyr and 25.6 kyr that rise above the 95% confidence interval of the red noise. The question is: what do those spectral peaks represent?

Tuning to offset the effects of variations in sediment accumulation rate and minor hiatuses throughout the section can be one way of addressing this question. In tuning, an obvious periodicity in the time series is aligned with a reference time series to make sure the maximum (or minimum) of the cycles are spaced evenly. If the section is young enough (less than 40–50 Ma) the time series can be tied to a theoretical time series for a given Milankovitch period (Laskar *et al.* 2004). Kodama *et al.* (2010) tuned the ARM time series of the Arguis Formation three times (once to 405 kyr long eccentricity cycles and in two iterations to 100 kyr short eccentricity). The first tuning to long eccentricity (Fig. 8.19) resulted in a very strong peak at 400 kyr, as would be expected. The check used on the viability of the tuning procedure is to see whether other Milankovitch frequencies are sharpened or emerge with the tuning. The long eccentricity tuning gives >99% confidence peaks at 250 kyr, 129 kyr, 23.8 kyr, 21.7 kyr, 19.7 kyr and 17 kyr. The 129 kyr peak is most likely short eccentricity and the *c.* 20 kyr peaks are most likely precession; the tuning is therefore validated in a sense, but these peaks were already evident in the untuned power spectrum.

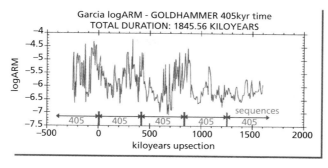

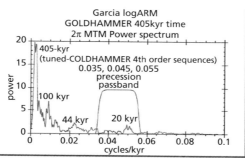

Fig. 8.17 The Cupido ARM cyclostratigraphy is tuned by tying the fourth-order sequence boundaries to long eccentricity (405 kyr periods). The time series analysis of the resulting ARM time series shows power at other Milankovitch frequencies (precession: 20 kyr; obliquity: 44 kyr; and short eccentricity: 100 kyr). From LA Hinnov, D Anastasio, D Latta, K Kodama and M Elrick, Milankovitch-controlled paleoclimate signal recorded by rock magnetics, Lower Cretaceous Platform carbonates of northern Mexico, American Association of Petroleum Geologists, Search and Discovery Article #40388, 2009, AAPG. The range of expected precession frequencies (0.035, 0.045 and 0.055 cycles/kyr) are shown in the bottom plot. (See Colour Plate 23)

Further tuning at short eccentricity by filtering the ARM time series at the tuning frequency for better comparison to the theoretical time series of course sharpens the short eccentricity peak, but significantly reduces the c. 200 kyr peak to below 99% confidence level for the red noise. The precession peaks stay strong. This would suggest that the 200 kyr peak that was so strong in the untuned ARM time series may have been an artifact of varying sediment accumulation rates and/or hiatuses in the section, although it should not be entirely dismissed as representing a real, but non-Milankovitch, cycle. The ARM record therefore provides good evidence that rock magnetic cyclostratigraphy can give a high-resolution chronostratigraphy, at the precessional timescale, for these growth strata.

An even more detailed study in the form of a coherency analysis, that compares the phase of the tuned ARM series to the phase of the precession-scale insolation, suggests that the precession peaks are in phase with autumn insolation peaks. The phase of precession insolation changes by 180° between June and December. Since the record was tuned to eccentricity, precession's phase should not be affected by the tuning. This observation suggests to Kodama *et al.* (2010) that the ARM is climate encoded by continental runoff that peaks during the fall rainy season in the Mediterranean paleoclimate for the Arguis Formation in the Eocene. This demonstrates the rich amount of information to be gleaned from detailed cyclostratigraphic analysis of good rock magnetic cyclostratigraphic records.

THE LATEMAR CONTROVERSY

The last example of rock magnetic cyclostratigraphy in this section is the so-called Latemar controversy. The basis for the controversy is simply that different geochronological techniques give different results for the duration of the deposition of a thick platform

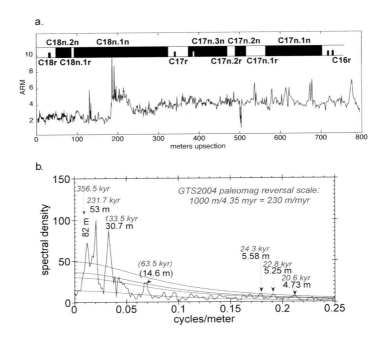

Fig. 8.18 Time series analysis of Arguis Formation ARM cyclostratigraphy without any tuning, simply tied to absolute time with magnetostratigraphy, shows spectral peaks at expected Milankovitch frequencies. Figure from Kodama *et al.* (2009). (See Colour Plate 24)

carbonate sequence (670 m) in the Triassic rocks in the Dolomites of northern Italy. U–Pb radiometric dating of zircons from ash layers in the Latemar sequence give ages very close to each other, suggesting the whole sequence was deposited in 2 million years (Mundil *et al.* 2003). The sequence contains *c.* 1 m thick upward shallowing sequences often observed in platform carbonates. A cyclostratigraphic analysis of these layers suggests that they are bundled in 5:1 packages (Goldhammer *et al.* 1990). The bundling is highly suggestive of short eccentricity to precession bundling and a Milankovitch forcing for the stratigraphy. However, this interpretation means that the Latemar had to have been deposited over nearly 12 million years, an order of magnitude difference with U–Pb dating results. Resolution of the controversy clearly has important implications for the validity of orbitally forced interpretations of stratigraphic cycles, particularly the 5:1 bundling often used to justify an astronomically forced interpretation. One additional piece of evidence

bearing on the depositional duration of the Latemar is a magnetostratigraphic analysis of the main Latemar sequence (Kent *et al.* 2004) observing only one polarity in the whole 670 m sequence, again suggesting deposition over only 1 million years or less. However, the magnetostratigraphy of these exposed rocks at high elevations in the Dolomites was plagued by lightning strikes that remagnetized a good portion of the dataset by application of natural IRMs, thus calling into question the validity of the magnetostratigraphy.

A rock magnetic cyclostratigraphy for a short sequence (40 m with c. 25 cm spacing) of Latemar equivalent rocks at Cimon Forcellone (Kodama & Hinnov 2007) was measured to see if rock magnetics would identify the same cyclicities observed in the lithologic cyclostratigraphy. This study was followed up by a larger rock magnetic study of the main Latemar sequence (100 m with *c.* 10–20 cm spacing; Marangon 2011). Both studies used ARM as a measure of magnetite concentration variations. Rock magnetic meas-

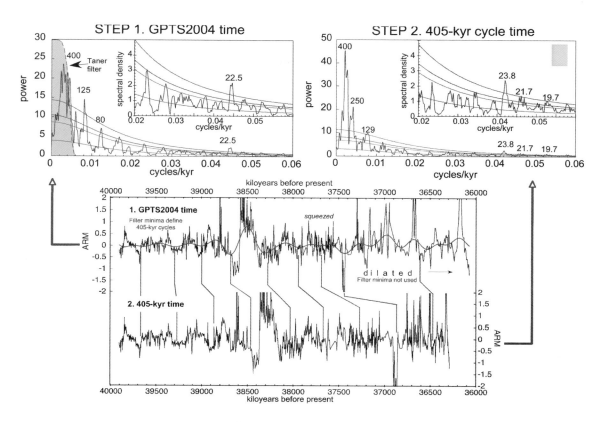

Fig. 8.19 An example of tuning to long eccentricity for the Arguis Formation (Kodama *et al.* 2009, 2010). In the upper left (Step 1) the power spectrum for the ARM cyclostratigraphy tied to magnetostratigraphy (GPTS 2004) shows the band-pass filtering at 400 kyr in gray shading. The red line in the center figure is the filtered ARM time series plotted on top of the unfiltered ARM time series. The filtered ARM time series is then tuned to even 405 kyr intervals giving the tuned series in the bottom of the center figure. The power spectrum for the 405 kyr tuned cyclostratigraphy is shown in the upper right (Step 2). (See Colour Plate 25)

urements clearly identified magnetite as the dominant magnetic mineral and both studies identified 1 m periodicity in the MTM analysis of the data that was bundled at 5 m (Fig. 8.20).

For the longer 100 m section at Cimon Latemar these two peaks rise above the 99% confidence level of the robust red noise model. Depth rank analysis of the lithologic facies for both sections also saw a 1 m periodicity bundled at 5 m, again significantly above the red noise for the 100 m long Cimon Latemar section. The ARM/SIRM ratio for the Forcellone samples showed magnetic particle sizes consistent with eolian dust for

the source of the magnetite. A simple orbitally forced interpretation of the rock magnetic data is that eolian dust, forced by global aridity beating to precession modulated by short-eccentricity Milankovitch cycles, was deposited in the Latemar carbonate platform in the Triassic. Most importantly for the cyclostratigraphic interpretation, 1 m of carbonate was deposited over about 20 kyr and 5 m in 100 kyr. This interpretation forces the 600+ m thick Latemar sequence to be deposited over 12 million years. It also suggests that sea level (as recorded by facies depth rank) variations are beating at the same periodicities as global aridity.

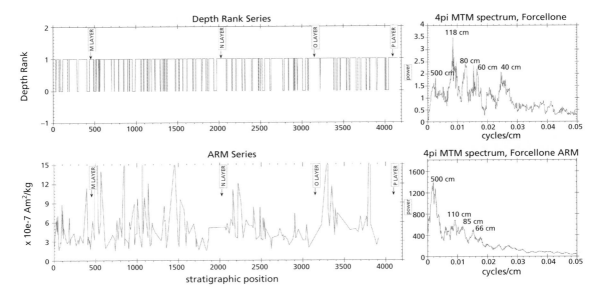

Fig. 8.20 Depth rank series and ARM cyclostratigraphy for the Forcellone section of the Latemar dolomites. The power spectra of these series show periodicities at 100 cm (118 and 110 cm) bundled at 500 cm, giving the 5 : 1 ratio expected for the Milankovitch frequencies of short eccentricity and precession. Figure from Kodama & Hinnov (2007). (See Colour Plate 26)

However, the rock magnetic and lithologic cyclostratigraphy is clearly at odds with the U–Pb and magnetostratigraphic interpretations. The magnetostratigraphy collected from the main Latemar sequence was compromised by remagnetization caused by lightning strikes, so another magnetostratigraphic section was sampled at Rio Sacuz (Spahn 2011; Spahn et al. 2011). The Rio Sacuz rocks are equivalent in age to the Cimon Latemar carbonate platform rocks, but Rio Sacuz was off the carbonate platform. Stratigraphic equivalence is shown by the identification of the ash layers at Rio Sacuz that can be tied to the ash layers in the main carbonate platform sequence at Latemar and Forcellone (Preto et al. 2007). The Rio Sacuz section is however in a protected stream valley below the tree line and was not affected by lightning strikes. The magnetostratigraphy for this section identified more than one polarity for the Latemar, a N-R-N-R sequence that, when tied to an integrated geomagnetic polarity timescale for the Triassic (Hounslow and Muttoni, 2010), indicates that the 670 m of the Latemar platform carbonate sequence was deposited over only about 1–2 million years. This is in good agreement with the U–Pb

geochronology and the previously reported magnetostratigraphy (Kent et al. 2004).

This result has important implications for rock magnetic cyclostratigraphic interpretations. It suggests that precession, if recorded by the rock magnetics, occurs at a 5 m period rather than at a 1 m period. What should then be made of the 5 : 1 m bundling touted as evidence for eccentricity: precession periodicities and an astronomically forced interpretation? If the 1 m upward shallowing sequences are the results of autocycles, why are they bundled at 5 m? Is it just a fortuitous juxtaposition of autocycles and precession? While this last point may be what happened in reality at the Latemar, it is hardly a satisfying result. What it does show is the importance of having independent absolute time control to bolster a cyclostratgraphic interpretation. Absolute time control can come from sequence stratigraphy, magnetostratigraphy, biostratigraphy, geochronology or some other well-accepted technique. It is apparent from the Latemar results that 5 : 1 bundling can fool workers into an astronomically forced interpretation; it may be that 5 : 1 bundling has a deeper meaning than simply astronomical forcing.

ROCK MAGNETIC CYCLOSTRATIGRAPHY: SUMMARY

These examples of rock magnetic cyclostratigraphy studies of the Eocene Arguis Formation, the Cretaceous Cupido Formation and the Traissic Latemar dolomites, as well as the striking 5 m periodicity of the goethite to hematite ratio in the Neoproterozoic Johnnie Formation (Figure 8.1), indicate that rock magnetic variations can detect subtle environmental changes which can be linked to astronomically forced global climate cycles. The technique has a promising future for providing high-resolution time control for sedimentary sequences, even as far back as the Late Precambrian.

The Magnetization of Sedimentary Rocks: Processes and their Interpretation

This last chapter is designed to bring together most of the major points from the book about processes that can affect the magnetization of a sedimentary rock, and to give an outline of the procedures that will, hopefully, ensure the best paleomagnetic results from a sedimentary rock. There is no guarantee that these techniques will always yield great results. As can be seen from Chapter 6 on reduction diagenesis and remagnetization, some sedimentary rocks have completely lost their primary magnetization. The experimental techniques outlined in this chapter will help the worker determine when this may have happened and incorporate that reality into their interpretation of the results. Of course, as in any endeavor, a prescribed set of procedures will be rigid and give a false sense of security of incontrovertibly robust results. We need to be creative in the design of a measurement procedure and a sampling scheme to meet the objective of the investigation. Knowing something about the different processes that could affect the accuracy of a paleomagnetic result will help in the design of the field and laboratory work so that many of the pitfalls in interpreting paleomagnetic data can be minimized and the best results obtained.

FIELD WORK: ROCK TYPE AND SAMPLING SCHEME

Practiced paleomagnetists can look at an outcrop of a sedimentary rock and immediately have a sense of whether it is likely to yield good results. This can only come from years of experience trying many different rock types; even with that experience, however, a good paleomagnetist can be fooled. Sometimes rocks that are not expected to give good results do; other times rocks that typically yield good results do not. Knowing something about how a sedimentary rock is magnetized and how a primary magnetization could be affected after it has been acquired can be a guide in the field, so that time is spent sampling the rocks that will most probably yield good results.

The best sedimentary rocks for paleomagnetism are typically fine grained. Mudrocks, siltstones or very fine-grained sandstones are the best for good paleomagnetic results because these grain sizes indicate a quiet depositional environment relatively free from the misaligning influences of currents and waves in a high-energy environment. Lake, hemipelagic marine, deep marine or fluvial overbank environments ensure rela-

Paleomagnetism of Sedimentary Rocks: Process and Interpretation, First Edition. Kenneth P. Kodama.
© 2012 Kenneth P. Kodama. Published 2012 by Blackwell Publishing Ltd.

tively quiet depositional environments and relatively continuous sedimentation, so that a complete record of the geomagnetic field is probable. Gray-colored mudrocks would usually suggest enough magnetite for a strong paleomagnetic signal; their fine grain size would increase the probability that it is well aligned with the ancient geomagnetic field.

Red sedimentary rocks, red shales, mudstones, silt-stones or sandstones are always good targets for paleo-magnetic studies because the red color indicates that hematite, a very stable magnetic mineral because of its high coercivity, is more than likely present in the rock. Of course, as indicated earlier in the book, the age of a red bed's magnetization may not be the same as its depositional age, but this can be investigated using different laboratory and field tests. The magnetization will most likely be stable and ancient.

Pure quartz sandstones (white in color) will probably have weak paleomagnetism and be more prone to secondary magnetic overprinting. This is also true for pure limestones, mainly because of the dearth of depositional magnetite in these rocks. Organic-rich lake and marine sediments and sedimentary rocks are likely affected by reduction diagenesis. The original depositional magnetite in these rocks will have been (at least partially and maybe completely) dissolved and replaced by secondary iron sulfides. However, if the sediment accumulation rate was high enough, the depositional magnetite may have passed through the reduction diagenesis zone fast enough such that little alteration occurred. These types of rocks should therefore not be avoided, but simply tested for the predominance of iron sulfide magnetizations.

Orange-brown weathering stains on the outcrop indicate present-day chemical weathering of the rock's iron minerals that has probably either altered the depositional Fe oxides or produced secondary magnetic minerals from the oxidation of iron silicates. Similarly, a sickly brown-yellow color for a fine-grained sandstone does not bode well for good paleomagnetic results. The color of the rock hints that secondary iron oxide hydroxides (e.g. limonite, goethite) have overprinted a weak primary magnetization.

Intensely deformed rocks, for instance sedimentary rocks with a well-developed spaced cleavage, will probably not yield pristine primary paleomagnetic results. They are likely to have secondary magnetizations formed during deformation or primary magnetizations that have been deflected by grain-scale strain. Metamorphic rocks of greenschist grade and higher were almost certainly reset paleomagnetically, either thermally or by the growth of secondary magnetic minerals from metamorphic fluids. Even lower-grade rocks such as slates probably had their primary magnetizations affected by deformation.

The best sampling strategy for a study depends on the objective of the study. Sampling schemes are as varied as the types of observations being made. For good paleomagnetic poles, either for global or regional tectonic studies, paleomagnetic sites should be distributed throughout a geologic formation to make sure changes in the geomagnetic field due to paleosecular variation are adequately averaged. Paleosecular variation is averaged over periods of 10^3–10^4 years, so even most sites in a sedimentary rock will average secular variation adequately. Recent plate motions suggest that sampling a unit that was deposited over periods of longer than 5 million years could include long-term directional changes due to plate motion. The unit should therefore be sampled over thicknesses that record shorter time periods than 5 or more million years. Although the definition of a paleomagnetic site is usually given as sampling the geomagnetic field at a geologic instant of time (e.g. Butler 1992), for paleomagnetic poles paleomagnetists typically collect about 8–10 individually oriented samples over a stratigraphic thickness of about a meter or less. This is done to ensure the best results as any one stratigraphic layer may not be a perfect paleomagnetic recorder; sampling over different strata at a site increases the chances of getting good results.

For magnetostratigraphic studies designed to catch the reversals of the geomagnetic field, sites are taken from as close to one stratum as possible so a geologic instant of time is captured. Many sites are however collected throughout the formation to see how the geomagnetic field is changing through time. Obviously, continuous sampling of the strata would be perfect but is not logistically possible. In order to properly identify a polarity interval, it is optimal to have at least three sequential sites of the same polarity. The stratigraphic sampling interval therefore depends on the expected reversal rate of the geomagnetic field and an estimate of the sediment accumulation rate for the rocks. The geomagnetic field's reversal rate is not stationary but has changed through geologic time, so a rough estimate of the time period sampled is very helpful to design a good sampling strategy. For instance, for the Mid-Tertiary the field reversed on average 4–5 times per million years. To get three sequential sites during a polarity interval in continental sedimentary rocks with an average sedimentation rate of 100 m/million years would require a stratigraphic sampling interval of about 5–10 m. However, for much more slowly deposited marine sediments with a sediment accumulation rate of 10 m per million years or less, the sites must be spaced at least every 75 cm. At the other end of the timescale for a magnetostratigraphic study, at least 10–20 polarity intervals need to be observed for a good fit to the geomagnetic polarity time scale (Butler 1992), so at least 2 million years of rocks should be sampled. Because many more sites are collected in a magnetostratigraphic study compared to a paleomagnetic pole study, only 3 or 4 samples are collected per site.

For secular variation studies of the geomagnetic field, lake or marine sediments are sampled nearly continuously in a core. The fine-grained sediments measured usually ensure a fairly continuous sedimentation rate, important for observing in detail changes in the geomagnetic field direction and intensity.

Field tests for the stability and age of the remanence are always critical to a good paleomagnetic study so sampling the limbs of folds, or at least strata with different tectonic tilts, is made whenever possible. Sampling the baked zone around an igneous intrusion and the unbaked rocks further away from the intrusion for a baked contact test, or sampling the clasts and matrix of a conglomerate layer for a conglomerate test, are also important if the opportunity arises.

STANDARD LABORATORY MEASUREMENTS AND ANALYSIS

Standard demagnetization techniques are used to remove secondary magnetizations acquired by rocks. In either of the two widely used techniques – alternating field demagnetization and thermal demagnetization – the magnetization of a rock is removed incrementally either by increasingly higher alternating magnetic fields or increasingly higher temperatures. The magnetizations for subpopulations of magnetic grains in the rocks with either coercivities or unblocking temperatures affected by the demagnetization fields or temperatures are essentially randomized by the procedure and no longer contribute to the magnetization of the rock when it is measured at each of the demagnetization steps. Both Butler (1992) and Tauxe (2010) provide excellent detailed treatments of progressive demagnetization; the reader is referred to these books for the theoretical and practical details of each type of demagnetization. The underlying assumption of demagnetization is that secondary magnetizations acquired by sedimentary rocks since the time of their deposition have either low coercivities or low unblocking temperatures, so progressive demagnetization is designed to isolate the magnetization direction of the oldest, primary magnetization of a rock. This assumption is not always borne out and needs to be rigorously tested with field tests that constrain the age of a rock's magnetization, i.e. fold tests, baked contact tests or conglomerate tests.

When detailed progressive demagnetization is applied to a rock, typically for ancient sedimentary rocks, multiple components of magnetization are removed at different ranges of unblocking temperatures or alternating fields (Fig. 9.1). The smaller the incremental steps of demagnetization, the easier it is to resolve the different components of magnetization in a rock. It is often advisable to be conservative in designing a demagnetization strategy and space demagnetization steps relatively closely together because, once a rock has been demagnetized at a particular field or temperature, that magnetization has been removed and the rock can never be measured again at lower steps. This last statement is perhaps obvious but it cannot be overstated: once a magnetization has been demagnetized, it's gone. When demagnetizing a rock it is particularly important to isolate the highest coercivity or unblocking-temperature magnetization component in a rock, i.e. the component that is removed just before the rock is completely demagnetized or the Curie temperature is reached. This is the most stable magnetization in a rock and is most likely to be the most ancient, perhaps the primary magnetization. Because it is not necessarily the primary magnetization, however, the last component of magnetization removed is denoted the characteristic remanent magnetization (ChRM).

A robust paleomagnetic result depends on the measurement of many samples from many sites from a sedimentary rock unit. Typically 8–10 samples are collected at each paleomagnetic site (less for a magnetostratigraphic study), and at least 15–20 sites are collected from a sedimentary formation for a reasonably well-constrained paleomagnetic pole. Many more sites are collected for a magnetostratigraphic study. There is always a tradeoff between minimizing the confidence limits around the mean paleomagnetic direction obtained by measurement and demagnetization of the multiple samples from a site and the multi-

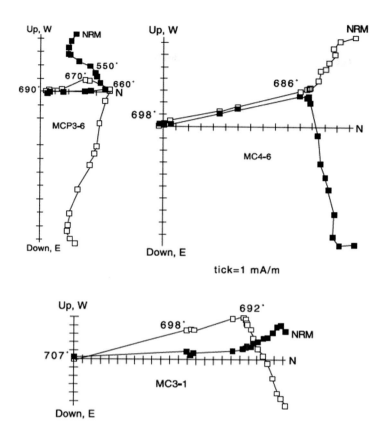

Fig. 9.1 Example of multi-component magnetizations in the Mississippian Mauch Chunk Formation red beds from northeastern Pennsylvania (Tan & Kodama 2002). These thermal demagnetization data are plotted in vector endpoint diagrams typically used to represent demagnetization data (Zijderveld 1967). The solid squares indicate the endpoints of the horizontal component of the paleomagnetic vector at each demagnetization step; the open squares indicate the endpoints of the vertical magnetization component. An easterly and steeply upward component is removed at temperatures up to 686°C for sample MC4-6 and then the characteristic magnetization is removed between temperatures of 686°C and up to 707°C. The characteristic remanence for this sample is directed northwesterly and upward at low inclinations. Butler (1992) presents a detailed and very clear explanation of these vector endpoint diagrams. X Tan and KP Kodama, Magnetic anisotropy and paleomagnetic inclination shallowing in red beds: Evidence from the Mississippian Mauch Chunk Formation, Pennsylvania, *Journal of Geophysical Research*, 107, B11, doi: 10/1029/2001JB001636, 2002. Copyright 2002 American Geophysical Union. Reproduced by permission of American Geophysical Union.

ple site mean directions from a formation, and the amount of work entailed in collecting, measuring and demagnetizing many samples.

Using Fisher (1953) statistics, Butler (1992) has shown that to average paleomagnetic directions and estimate the 95% confidence limits of a mean direction, the confidence limits do not decrease appreciably after about 20 directions are averaged. Fisher statistics

are the traditional mathematical method used to average paleomagnetic directions, but these statistics assume a particular directional distribution of directions (i.e. a Fisher distribution that is a two-dimensional unit vector distribution on the surface of a sphere, either the Earth or a stereonet representation of the data). The 95% confidence limits tell you that there is 95% probability that the true mean, were you to measure an infinite number of samples, lies within the 95% confidence limits of the mean obtained by your markedly smaller dataset. The smaller the confidence limits the better but, whatever their size, they quantify at what level of confidence a paleomagnetic result can be used to determine a vertical axis rotation, the position of a plate or continent in the past or the variations of the secular variation of the geomagnetic field.

Tauxe (2010) has pointed out that many directional distributions measured for paleomagnetic samples are non-Fisherian. She has championed and pioneered the statistical analysis of paleomagnetic directions using bootstrap techniques that can be designed so they do not make any assumptions about the directional distribution of the paleomagnetic data. In bootstrap analysis, the observed data distribution is resampled thousands of times by a computer randomly selecting different points from the original dataset. An estimate of the mean direction and its confidence limits are derived from the scatter of the means calculated from the multiple resampling of the dataset. In order for the technique to work well, the initial dataset needs to be large enough to withstand the multiple resamplings.

One way that paleomagnetic directional distributions are non-Fisherian is when they are elliptical. Elliptical distributions are better statistically analyzed with the Kent (1982) distribution that is the elliptical analogue of the Fisher distribution. However, both Fisher and Kent distributions assume that the directional distributions are one polarity. Obviously paleomagnetic directional distributions come in two flavors of reversed and normal polarity. Most geologists handle the statistical analysis of bi-modal distributions by inverting one of the polarity distributions through the origin to create a distribution of one polarity. There is always some ambiguity in this approach however, particularly if the results are scattered, since it is hard to unequivocally determine the polarity of a very divergent direction.

For these types of distributions, it is best to use Bingham (1974) statistics that can handle bi-modal direction distributions. However, Bingham statistics assume that the two modes are antipodal, i.e. exactly opposite to each other in direction; because of overprinting or inadequate sampling of secular variation, this is not always the case.

There are therefore advantages and drawbacks for using each type of statistics for analysis. Fisher and Bingham statistics are both widely used in paleomag-

netic data analysis. Recently, bootstrapping of paleomagnetic data has been gaining ground. Tauxe (2010) gives more detailed coverage of different kinds of paleomagnetic data statistical analysis.

Most workers have found that thermal demagnetization does a much better job than alternating field demagnetization in removing the multiple secondary magnetizations in ancient (early Tertiary, Mesozoic, Paleozoic, or Precambrian) sedimentary rocks and so, for most modern paleomagnetism studies, it is the demagnetization technique of choice. Alternating field demagnetization works well on young lake and marine sediments that are difficult to heat because of their high organic or water content. Alternating field demagnetization also works well on young glacial sediments that usually contain freshly eroded magnetite liberated by intense mechanical weathering and not appreciably affected by chemical alteration.

There are two common cases when thermal demagnetization is problematic. If a sedimentary rock contains secondary iron sulfides, formed by reduction diagenesis, the sulfides often oxidize to magnetite at heating near to the Curie Point of the iron sulfides ($c.$ 300°C). The magnetite formed is usually unstable magnetically and its magnetization masks the magnetization of any stable, depositional magnetite present in the rock that has higher unblocking temperatures than the iron sulfides, making it difficult to isolate the initial depositional remanence carried by the primary magnetite. Therefore, simple heating of an iron-sulfide-bearing rock above the 300°C sulfide Curie temperature does not necessarily isolate the magnetization of the depositional magnetite in a rock. The instability of the secondary magnetite formed during heating is probably because it is very fine grained and superparamagnetic.

The other case where thermal demagnetization does not work particularly well is the removal of secondary overprinting caused by lightning strikes. Lightning strikes and the resulting huge currents through the ground produce correspondingly very high magnetic fields and strong natural IRMs. IRMs are difficult to remove by thermal demagnetization, and so it is best to avoid sampling localities struck by lightning. High exposed ridges are particularly prone to lightning strikes. Sometimes slowly sweeping a field compass along the ground near a potential sampling site and watching for large deflections of the compass needle is a good way to identify, and avoid, places probably struck by lightning. Since magnetic fields fall off in intensity by the inverse cube of distance, lightning effects tend to be very localized.

There are other demagnetization techniques that are much less commonly used, but are sometimes good for attacking a particular problem. Chemical demagnetization entails soaking paleomagnetic samples in a concentrated acid (usually HCl) for increasingly longer periods of time (up to several months) and measuring them at increments of increasingly long duration. The underlying assumption is that the finest magnetic grains are dissolved first by the acid soaking into the rock, leaving the largest magnetic grains at the longest periods of chemical demagnetization. This assumption is reasonable for hematite-bearing red beds because hematite, the magnetic mineral of red beds, has a relatively large SD-MD critical diameter close to *c.* 10–20 μm (Dunlop & Ozdemir 1997). The submicron pigmentary hematite grains that give red beds their color are assumed to be secondary and should be removed at earlier steps of chemical demagnetization while the larger micron-sized supposedly detrital hematite grains are removed last. This assumption may not always be borne out and should be checked by thermal demagnetization that works quite well for hematite-bearing red beds. Because chemical demagnetization is messy and takes a very long time to complete (months) it is no longer widely used. It is also not always as successful as thermal demagnetization in isolating a rock's primary magnetization. Tan & Kodama (2002) and Tan *et al.* (2003) used chemical demagnetization to isolate the AMS of the hematite grains carrying the characteristic magnetization (ChRM) of red beds for inclination shallowing studies.

Microwave demagnetization can be used to preferentially heat the magnetic minerals in a rock without appreciably heating the whole specimen and causing unwanted mineralogic changes. Samples are exposed to microwave radiation in a resonant cavity for several seconds in microwave demagnetization. The microwave radiation affects the spin moment alignment of the Fe atom electrons that generate the spontaneous magnetization of a magnetic mineral. The technique is more typically used to apply remanences to rocks for absolute paleointensity measurements. In order to demagnetize a sample with microwaves, some energy is passed to the surrounding non-magnetic minerals and heats the specimen to temperatures of 150°C or less (Shaw 2007).

In laser selective demagnetization (LSD; Renne & Onstott 1988) magnetic minerals are selectively removed in thin section by laser pulses, thus allowing the removal of specific magnetic phases and the determination of what subpopulation of magnetic grains carries what component of remanence. This can be useful for unraveling the relative ages of multi-component magnetizations in a rock, since cross-cutting relationships between different magnetic phases can be observed by microscopic examination of thin sections.

Once progressive demagnetization has been completed for a collection of samples from a sedimentary rock and the characteristic magnetization of the rock has been isolated, several laboratory measurements need to be made to check the age, the source and the accuracy of the characteristic remanence. These tests are the focus of this chapter.

CONSTRAINING THE AGE OF MAGNETIZATION

Graham's fold test is an important field test that paleomagnetists use to constrain the age of magnetization in a rock. After demagnetization is completed and the characteristic remanent magnetization (ChRM) has been isolated, it is crucial to test whether the magnetization is in fact primary or at least nearly the same age as the age of the sedimentary rock. The fold test has been discussed already in Chapter 7 as grain-scale rock strain can affect the interpretation of the fold test, particularly the syn-folding magnetization that occurs when the paleomagnetic directions on either limb of a fold cluster best when the bedding tilts are only partially removed during the test. Basically, the fold test is conducted by sampling both limbs of a fold or regionally sampling beds with different bedding orientations. If the magnetization clusters better in 'untilted' coordinates when the limbs have been restored to their horizontal pre-folding orientation, the magnetization was acquired before folding occurred. If the magnetization clusters better before the beds are 'untilted', the magnetization was acquired some time after tilting of the beds. Chapter 7 considers in detail the case when the magnetization clusters best about half-way in between. The most straightforward interpretation is of course that the magnetization was acquired during folding, but strain can have an effect.

It is of course important to quantitatively assess at what level of confidence the magnetization has passed or failed the fold test. For this assessment, statistics are needed to determine whether the clustering in a given orientation is better than in another configuration. Many different statistical tests have been proposed and used (e.g. McElhinny 1964; McFadden & Jones 1981; McFadden 1990; Tauxe & Watson 1994; Enkin 2003). Both Butler (1992) and Tauxe (2010) provide good descriptions of these tests and how to use them, but they all have one result in common: they indicate that at some level of confidence, usually 95% in most paleomagnetic studies, whether the best clustered magnetization is significantly different in its clustering than in its worst clustered configuration. For the fold test to be meaningful, it is helpful if the folding occurred geologically soon after the sedimentary rocks were deposited.

A conglomerate test constrains the age of magnetization by comparing the magnetization in conglomerate clasts with the magnetization in the matrix. The magnetic directions in the clasts should be randomly oriented if the rock has not been affected by a pervasive remagnetization. Statistical tests for random directions are used to quantify the conglomerate test (Watson 1956). The conglomerate test is not used as frequently as a fold test, probably because conglomerate layers are not particularly good paleomagnetic targets because they indicate a high-energy environment of deposition.

A baked contact test checks if the rocks heated by an igneous intrusion have been thermally remagnetized in a direction similar to that of the igneous rocks and different from the surrounding unheated rocks. This configuration of magnetizations would suggest that the magnetization in the surrounding unheated rocks (and of course, the baked rocks and the igneous rocks) is not the result of a large-scale remagnetization and is, therefore, assumed to be primary. Although it is an important test, it does not constrain the age of magnetization as definitively as a fold test. The test can be quantified by using statistics to check whether the mean directions of the different rock units are the same or different at a given level of confidence, typically 95%. McFadden & Lowes (1981) have provided a discrimination of means statistical test that is typically used.

Workers often support an ancient age of magnetization by comparing the characteristic magnetization isolated by demagnetization to the present-day field direction, the axial dipole field direction during the most recent normal polarity chron (Brunhes chron) or to an ancient, but younger, paleomagnetic field direction from the tectonic unit being studied (e.g. continent, plate). Although these comparisons are not as powerful as the three field tests described above, they provide additional evidence for or against a remagnetization in either the present-day field, the most recent normal polarity chron or at some time after deposition of the sedimentary rocks.

Finally, the magnetic fabric of a sedimentary rock can be used to argue whether or not the rock is carrying a primary depositional remanence. This test is not generally used, but could become important if magnetic fabric were more routinely measured, for example, to check for inclination shallowing. It is not as powerful as the fold test nor can it be quantified by statistics, but the observation of depositional/compactional magnetic fabric – particularly a remanence fabric carried by the same magnetic grains that carry the ChRM, with minimum principal axes perpendicular to bedding and maximum and intermediate axes scattered in the bedding plane – is a strong argument that the magnetization is either depositional or acquired very soon after deposition before much burial compaction has occurred.

SOURCE OF A ROCK'S MAGNETIZATION: MAGNETIC MINERALOGY

It is incredibly important to identify the magnetic minerals carrying the magnetization of a sedimentary rock in order to gain insight into the source of the magnetization. Different magnetic minerals are likely to be a primary depositional magnetic mineral or a secondary magnetic mineral. Magnetite is the expected primary depositional mineral for marine, fluvial and lake sediments. It is transported to the depositional basin from either igneous, metamorphic or sedimentary rock source areas. The magnetite has probably been initially magnetized by thermal processes. In some cases, magnetic minerals that are usually thought to be secondary, e.g. maghemite or pyrrhotite, have been shown to carry a depositional remanence (Kodama 1982; Horng & Roberts 2006), but magnetite is the premier primary magnetic mineral. Hematite, particularly specular hematite, is the expected primary magnetic mineral in red sedimentary rocks, i.e. red beds. As described in Chapter 6, it is 'up in the air' whether the hematite in red beds is primary depositional or secondary; if it is

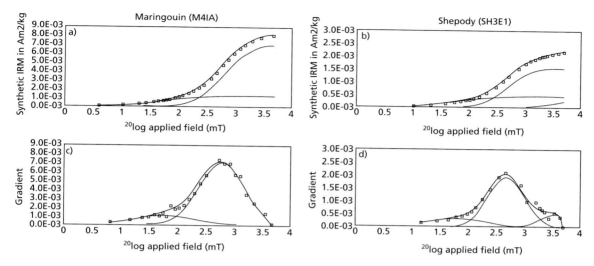

Fig. 9.2 Example of IRM acquisition modeling using the technique of Kruiver *et al.* (2001). Bilardello & Kodama (2009b) used IRM acquisition modeling to show that the Shepody and Maringouin red beds from the Canadian Maritimes had both high-coercivity hematite (coercivities peaking around log 2.7 mT or 500–630 mT) and lesser amounts of magnetite (coercivities peaking around log 1.5 mT or 50–60 mT). The top curves are standard IRM acquisition curves; the bottom curves are their first derivative which is used for the modeling. D Bilardello and KP Kodama, Palaeomagnetism and magnetic anisotropy of Carboniferous red beds from the Maritime Provinces of Canada: evidence for shallow palaeomagnetic inclinations and implications for North American apparent polar wander, *Geophysical Journal International*, 180, 1013–1029, 2010, John Wiley & Sons, Ltd.

secondary, it was probably formed soon after the deposition of the rocks. Magnetic iron sulfides, greigite and pyrrhotite indicate reduction diageneis and are secondary magnetic minerals, but they have probably formed soon after deposition (within thousands of years or less). Goethite is typically the product of present-day chemical weathering, but could have been formed by weathering at any point in the postdepositional history of a rock.

The following laboratory experiments are typically used to identify or at least constrain the magnetic mineralogy of a sedimentary rock. Their results should be interpreted in light of all the information available about a rock: its lithology, likely depositional environment, demagnetizaton behavior and the results of field tests to constrain the magnetization age.

Laboratory tests for magnetic mineral identification

In isothermal remanent magnetization (IRM) acquisition experiments, a rock sample is exposed to stronger and stronger DC magnetic fields with the remanence being measured after each exposure. Usually an impulse magnetizer that discharges a capacitor through a coil is used to impart the DC magnetic fields, and most impulse magnetizers can reach fields as high as 5 T. The field strength is increased until the rock magnetically saturates, i.e. even though the field strength is increased, the strength of magnetization no longer increases but 'levels off'. Kruiver *et al.* (2001) have developed a technique for mathematically modeling IRM acquisition curves (Fig. 9.2). The modeling assumes that subpopulations of magnetic grains in the sample have different coercivities. The coercivity distributions of the subpopulations all contribute to the reconstruction of the first derivative of the IRM acquisition curve by forward modeling. Using this modeling, the different coercivity components in a rock are identified and quantified. Low coercivities (<100 mT) are usually interpreted to indicate magnetite; higher coercivities indicate either hematite (hundreds of mT to 1 T) or goethite (>1–2 T).

The so-called Lowrie test (Lowrie 1990) uses both the coercivity of magnetic minerals and their Curie or

Neel temperatures, the temperature at which they lose their spontaneous magnetization, to identify the magnetic minerals in a rock. In a standard Lowrie test, three IRMs are applied to a rock sample in orthogonal directions. The highest-field IRM is always applied first, then the intermediate-field IRM followed by the lowest-field IRM. Typical fields used are 1 T, c. 0.5 T and 0.1 T, respectively, since they are the coercivities observed for hematite or goethite at the high end and magnetite at the low end. The intermediate-field IRM chosen depends on the magnetic minerals suspected in the rock or sediment. For example, pyrrhotite can sometimes have intermediate coercivities so a field of 0.5 or 0.6 T can be chosen. Magnetite has a theoretical maximum coercivity of .0.3 T for very long needle-shaped particles, so 0.3 T can be chosen in that case. The three-IRM rock sample is then thermally demagnetized to observe the unblocking-temperature behavior of the different IRMs in the sample. The combination of coercivity and unblocking temperature is a powerful identification tool. For instance, a 580°C loss of magnetization for the 0.1 T IRM is pretty clear evidence of magnetite, while a 680°C loss of magnetization for the 1 T IRM is good evidence for hematite. Fe sulfides, usually with low to intermediate coercivities, lose their magnetization at 300°C while goethite loses its high coercivity IRM (>1 T) at c. 120°C.

A very powerful way of identifying the magnetic minerals in a specimen and determining whether they are primary depositional minerals or secondary authigenic minerals is by simply looking at them. Examination of the minerals in situ would of course be preferable, but their concentration is so low and their size so small (submicron) that it is not always practical for routine analysis.

Suk et al. (1990) observed the secondary magnetite in situ from remagnetized Late Paleozoic carbonates from Tennessee with TEM and SEM, but this was the focus of a detailed and painstaking study of the Kiaman remagnetization in eastern North America. For more routine analysis as part of a standard paleomagnetic study of sedimentary rocks, magnetic extraction of the magnetic minerals in a sediment or rock is the best way to characterize the main magnetic minerals.

Hounslow & Maher (1996) have outlined various techniques in detail, but they usually involve circulating a slurry made from the disaggregated rock or from the sediment past a strong magnetic field gradient created by small intense magnets, usually rare Earth magnets (Fig. 9.3). It is often difficult to liberate and

Fig. 9.3 Example of a magnetic extraction set-up. Slurry from a pulverized red bed was centrifuged to produce the red-colored material in the device. Hematite particles were extracted and their individual particle anisotropy was measured (Kodama 2009). (See Colour Plate 27)

extract all the magnetic minerals in a rock and the strongest magnetic minerals (usually magnetite) are preferentially removed; these points must be kept in mind when interpreting the examination of the extract under an electron microscope.

A combination of centrifuging and magnetic extraction from a sediment slurry has been used to extract more weakly magnetic hematite particles from a red bed (Dekkers & Linssen 1991; Kodama 2009). The magnetic extract can be examined under SEM or TEM. More perfectly shaped crystals suggest authigenesis of secondary minerals; more rounded irregularly shaped grains suggest sedimentary transport and primary depositional minerals. Chains of perfect single-domain -sized magnetite or greigite crystals obviously indicate formation by magnetotactic bacteria. The magnetic extracts can also be examined by energy-dispersive spectrometry (EDS) or microprobe to more directly determine their chemical make-up, for example whether they contain sulfur, to aid in their identification. X-ray diffraction of the extract, if there is enough of it, can also be important to magnetic mineral identification.

To check the efficiency of the magnetic extraction, it is useful to compare a rock magnetic measurement (e.g. an IRM acquisition curve) for the whole rock and for a sample made up from the magnetic extract.

ACCURACY OF THE PALEOMAGNETIC DIRECTION: DEFLECTION DUE TO PHYSICAL PROCESSES

Chapter 4 shows that burial compaction and sometimes syn-depositional processes can cause the deflection of a sediment's or sedimentary rock's DRM or pDRM inclination toward the horizontal, i.e. inclination shallowing. Since magnetite is unquestionably a depositional magnetic mineral, magnetite-bearing lake and marine sediments and sedimentary rocks should be checked for the effects of inclination shallowing once the characteristic magnetization has been isolated by demagnetization and the characteristic remanence has been shown likely to be primary by field tests. The anisotropy of remanence applied to the characteristic remanence-carrying magnetic particles, usually with an ARM, is a relatively straightforward way to check for inclination shallowing and to correct it. If there are enough sites (>100) then the EI technique (Tauxe & Kent 2004) is another very good way to check for and correct any inclination shallowing in a magnetite-bearing rock.

Hematite-bearing red beds have also been shown to suffer from inclination shallowing; however, it is not always clear that the magnetization of red beds is a primary depositional remanence. Red bed remanence could have been acquired early enough after deposition, by an early CRM for example, that it was subsequently affected by burial compaction. The anisotropy of remanence technique, this time using an IRM applied to the high-coercivity hematite particles, or the EI technique can be used to correct any inclination shallowing found in the red beds.

For magnetite and hematite-bearing sedimentary rocks the magnitude of inclination shallowing effect is about the same, ranging c. 10–20°, enough to give maximum inaccuracies in paleolatitude by about the same order of magnitude.

A sedimentary rock's remanence can also be deflected by grain-scale tectonic strain. Theoretically the size of the deflection can be quite large, particularly if the magnetization behaves in simple shear strain as if it were carried by actively rotating rigid particles. If the magnetization behaves more as a passive line, then it can only rotate as far as the shear plane. For a flexural slip fold, the shear plane is the bedding plane so inclinations would not change their sign for passive line strain behavior. Most paleomagnetists would avoid sampling severely strained sedimentary rocks, so the size of tectonic deflection is more likely to be of the order c. 10° if it has occurred at all in rocks typically sampled for paleomagnetism.

One way to check for strain deflection of remanence is to compare the paleomagnetic directions in a rock with rock strain measured by standard structural geology techniques, such as R_f-ϕ or center-to-center, across a region or locale of varying strain. Another possible indicator of strain-induced remanence deflection is a syn-folding magnetization, although the simplest interpretation of a syn-folding magnetization is a secondary remanence that occurred during folding. The observation of a syn-folding magnetization should be followed up with rock strain measurements to check for the possibility of remanence deflection by strain.

As detailed in Chapter 6, chemical remanence that has resulted from remagnetization or partial remagnetization of a sedimentary rock is of course an important source of inaccuracy of a paleomagnetic result. The CRM resulting from a remagnetization is not inaccurate *per se* – it is probably a very faithful record of the geomagnetic field at the time of remagnetization – but the age of magnetization is quite different from the depositional age of the rock. Fold tests and other field tests can be used to check for secondary magnetizations. Examination of a rock's magnetic minerals, either *in situ* or in a magnetic extract, can determine whether they are secondary in nature from their composition or from their morphology.

Viscous remanent magnetization (VRM) is another source of inaccuracy in a paleomagnetic direction, if it contributes to the final demagnetized paleomagnetic direction. Viscous remanence is a secondary magnetization acquired by rock over time when it is exposed to a post-depositional magnetic field. The mechanism of VRM acquisition is discussed in detail by both Butler

(1992) and Tauxe (2010) but low-stability magnetic grains, either very small grains close to superparamagnetic in particle size or large multi-domain grains with many magnetic domains, are the most susceptible to acquiring a viscous remanence. Typically, thermal and/or alternating field demagnetization removes a VRM, but some VRMs are difficult to remove. Storage in low field space before and during demagnetization and measurement, more common in modern paleomagnetic studies that house measurement and demagnetization equipment in magnetically shielded rooms with fields of 500 nT or less, minimizes the contribution of VRMs to the final paleomagnetic result.

A reversals test, in which the anti-podality of reversed and normal polarity directions is checked statistically (e.g. McFadden & McElhinny 1990), can provide a hint of the importance of a VRM in a paleo-

magnetic result. In Tertiary rocks with normal polarity directions similar to the present-day field, the reversed-polarity direction is sometimes not as steep as the normal polarity direction; this suggests unremoved normal polarity overprinting, typically attributed to a VRM.

If a rock has been heated in its post-depositional history, VRM would have been acquired at a higher rate and the rock would have acquired a thermoviscous remanent magnetization (TVRM) overprint. Pullaiah *et al.* (1975) have established a theory for SD grains that indicates what thermal demagnetization temperatures are needed to remove a TVRM for the 60 minute heating usually applied in a laboratory thermal demagnetization. However, Kent (1985) conducted a beautiful test observing the 10,000 year old VRM in conglomerate clasts deposited in glacial sediments. The

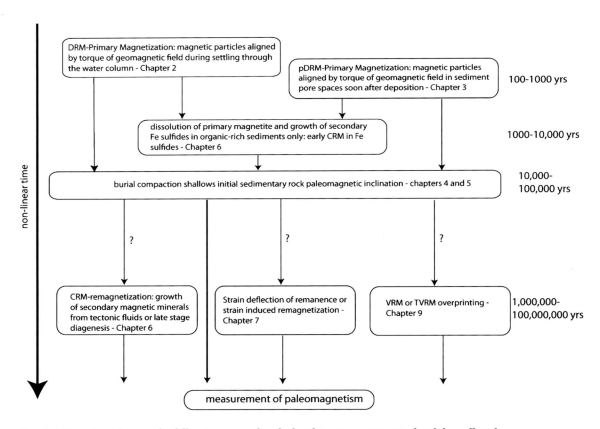

Fig. 9.4 Flow chart showing the different processes by which sediments are magnetized and then affected post-depositionally.

rocks Kent studied were clasts of the Paleozoic lime-stones that carry the Kiaman-age Late Paleozoic remagnetization discussed in Chapter 6. The clasts were remagnetized by a TVRM, with little heating, during the Holocene. Thermal demagnetization at temperatures higher than those predicted by Pullaiah *et al.* suggests that TVRM may be more pernicious than theory indicates.

SEDIMENTARY ROCK PALEOMAGNETISM: PROCESS AND INTERPRETATION

This book has been written to give geologists and students of paleomagnetism a sense of how sediments and sedimentary rocks acquire their paleomagnetism and the different processes that can affect a primary magnetization during the post-depositional history of a rock. It is written to give geologists a better appreciation of how paleomagnetists understand how a sedi-mentary rock is magnetized and what tests should be used to check the age and accuracy of the magnetiza-tion of sedimentary rock. The flowchart depicted by Fig. 9.4 summarizes the main processes covered in this book, including references to the chapter where the process is discussed and an estimate of the time after deposition when the process occurs.

The main point to take away from this book is that, although there are many processes that can affect the accuracy and age of the paleomagnetism of a sedimen-tary rock, there are many laboratory and field tests that can check for inaccuracies and the age of magnetiza-tion. Some laboratory experiments can correct the inaccuracies; other laboratory or field tests simply make the worker aware of processes that affected the magnetization of the rock subsequent to deposition so they can be considered in the interpretation of the paleomagnetic results. Many paleomagnetists already do this, but hopefully this book can act as a guide for those who want to evaluate paleomagnetic data in the literature or acquire their own paleomagnetic dataset.

Glossary of Paleomagnetic and Rock Magnetic Acronyms

AAR, anisotropy of anhysteretic remanence: The anisotropy that occurs when an ARM or a partial ARM is applied in a minimum of six different orientations to a sample, usually along the three orthogonal sample coordinate axes (X, Y and Z) and along the three orthogonal directions between each of these axes (XY, YZ and XZ). Some workers prefer to abbreviate this measurement as AARM for anisotropy of anhysteretic remanent magnetization.

AIR, anisotropy of isothermal remanence: The anisotropy that occurs when an IRM is applied in a minimum of six different orientations to a sample (i.e. X, Y, Z, XY, YZ and XZ). Often used to activate higher-coercivity minerals than those that can be activated by an ARM.

AMS, anisotropy of magnetic susceptibility: The anisotropy that results for an induced magnetization that is caused by application of a magnetic field in at least six different orientations to a sample. The magnetic field typically has the strength similar to the Earth's magnetic field, $c.$ 50 µT.

ARM, anhysteretic remanent magnetization: A laboratory remanence that results from applying a small DC magnetic field ($c.$ 50–100 µT) to a sample in the presence of an alternating magnetic field that is decreased from some peak value to 0. In most laboratories the peak field is $c.$ 100 mT. A partial ARM can be applied by only switching on the DC field over a limited range of alternating field during the decrease from a peak alternating field value.

B_c: The coercivity of a sample that results from a hysteresis loop measurement. It is the field that is required to reduce the magnetization of a sample to 0 while the field is being applied.

B_{cr}: The coercivity of remanence of a sample that results from a hysteresis loop measurement. It is the field required to reduce the magnetization of a sample to 0 after the field has been turned off, i.e. the sample is measured in 0 field.

CRM, chemical remanent magnetization: The remanent, or spontaneous, magnetization acquired when a ferromagnetic mineral grows chemically, or crystallizes, in the presence of a magnetic field. When new crystal grains grow through a certain volume, typically submicron in size, the spontaneous magnetization becomes stable.

ChRM, characteristic remanent magnetization: The stable magnetization that results from the complete demagnetization of a sample at the higher demagnetization fields or temperatures. If a sample has multiple components of magnetization revealed during demagnetization, the ChRM is the magnetization removed at the highest demagnetization field or temperature.

DRM, detrital remanent magnetization (or, to some, depositional remanent magnetization): The magnetization acquired by a sediment when its primary depositional magnetic minerals are aligned with the ambient magnetic field at deposition.

EI, elongation–inclination: Used for an inclination shallowing correction technique in which the elongation (ellipticity) of the directional scatter resulting from secular variation at a given latitude is matched to the mean inclination that would occur in a GAD field for that latitude. The EI pair used for the inclination correction is derived from the secular variation behavior of the present geomagnetic field over the past 5 million years.

FMR, ferromagnetic resonance: An electron spin resonance spectroscopic technique used to understand the magnetic structure of ferromagnetic materials, including the magnetic anisotropy of individual particles and magnetostatic interactions between magnetic particles.

FORC, first-order reversal curve: A curve that results from a complicated procedure that includes contouring of multiple hysteresis loops with decreasing peak fields from magnetic saturation down to nearly 0, and changing the coordinate system from that used for the generation of the hysteresis loops. The FORC diagram that results can be used to measure the coercivity of the collection of magnetic grains in a sample as well as the importance of magnetic interactions between the magnetic sample grains.

GAD, geomagnetic axial dipole: The geometric configuration of the Earth's magnetic field when secular variation has been time averaged. This is the field that results from a magnetic dipole at the center of the Earth that is parallel to the Earth's rotational axis.

IRM, isothermal remanent magnetization: The magnetization that results from the application of a DC magnetic field to a sample.

J_{sat}, saturation magnetization: This measurement results from a hysteresis loop measurement and is the greatest magnetization acquired by a sample in the presence of the magnetic field causing it.

J_{rs}, saturation remanence, also known as a SIRM: This is the greatest magnetization acquired by a sample after it has reached J_{sat}, but it is measured when the field that caused it has been turned off.

MTM, multi-taper method: A modification of Fourier spectral analysis in which multiple spectral estimates are calculated from the same sample, each time the data series being multiplied by a different orthogonal taper function. The technique was developed by DJ Thomson.

NRM, natural remanent magnetization: A blanket term used to indicate all remanent magnetizations that are naturally acquired by a rock. It usually refers to the magnetization measured before a rock sample is demagnetized or before it has had a laboratory remanence applied to it.

OAI, oxic–anoxic interface: The boundary between the oxygen-rich and oxygen-free depths, either in the water column or in the sediment column.

pDRM, post-depositional remanent magnetization: The DRM that a sediment acquires at some point after deposition. It is usually acquired within the top 10–20 cm of the sediment column, if it occurs at all.

PSD, pseudo-single domain: A small, multi-domain ferromagnetic grain that behaves like a single-domain grain. It is usually envisioned as containing just a few domains (regions of uniform magnetization), but PSD behavior could also be due to non-uniform magnetization configurations (vortex or flower) as indicated by micro-magnetic modeling.

PSV, paleosecular variation: The directional and intensity changes in the local geomagnetic field vector caused by the dynamic behavior of the Earth's magnetic field on timescales of hundreds–thousands of years. The dynamic behavior includes wobbling of the main dipole (c. 90% of the field) around the Earth's rotational axis, the growth and decay in the strength of the dipole, the movement of the non-dipole field foci (where field lines project in and out of the Earth) and the growth and decay of the non-dipole foci. The non-dipole field foci are like small N and S poles in the Earth's field after the main dipole has been mathematically subtracted. Their occurrence causes the geomagnetic field to deviate from a simple dipole in configuration.

SIRM, saturation isothermal remanent magnetization: The isothermal remanence acquired by a sample after it has been exposed to a saturating DC field. See J_{rs} above.

χ, susceptibility: The proportionality constant between the induced magnetization and the field that is inducing it ($J = \chi H$). For AMS, the susceptibility is described by a second-rank tensor (3×3 matrix).

χ_{fd}, frequency-dependent susceptibility: In a laboratory measurement of susceptibility, the inducing field used (H) is an alternating field so that no VRM is induced in the sample during the measurement. For χ_{fd} the susceptibility is measured in alternating fields with different frequencies, usually an order of magnitude or greater different. If the susceptibility varies greatly with frequency, it is a measure of superparamagnetic grains in a sample.

χ_{ARM}, ARM susceptibility: In this case not the susceptibility that causes an induced magnetization, but the ARM acquired as a function of the DC magnetic field used for the application of the ARM. Usually it is calculated by dividing the ARM by the strength of the DC field used during application of the ARM.

TRM, thermal remanent magnetization or thermoremament magnetization: The magnetization acquired by a ferromagnetic mineral grain during cooling of the grain through its Curie temperature or its blocking temperature. Blocking temperature is a function of magnetic grain size and is always less than the Curie temperature. It is the temperature below which the magnetization of the grain becomes stable for geologically significant lengths of time.

TVRM, thermoviscous remanent magnetization: A viscous magnetization (see VRM below) acquired by a ferromagnetic mineral grain at a given temperature for a given length of time, usually a temperature elevated above typical conditions in the laboratory.

VGP, virtual geomagnetic pole: The position of the north magnetic pole on the surface of the Earth that would generate the paleomagnetic direction observed for a rock sample, assuming that the Earth's field is caused by a dipole at the center of the Earth. Since PSV has not been time averaged, VGPs do not typically lie on the N geographic pole (spin axis of the Earth) but cluster around the N geographic pole during times of normal polarity.

VRM, viscous remanent magnetization: The magnetization acquired by a ferromagnetic mineral grain during exposure to a magnetic field, either in the laboratory or the Earth's magnetic field, over a given length of time. VRM is acquired as a function of the log of time in the laboratory. Its acquisition is greater at higher temperatures.

References

Addison, F.T., Turner, P. & Tarling, D.H., 1985. Magnetic studies of the Pendleside Limestone; evidence for remagnetization an dlate-diagenetic dolomitization during a post-Asbian normal event, *J. Geol. Soc. London*, 142, 983–994.

Ague, J.J. & Brandon, M.T., 1996. Regional tilt of the Mt. Stuart Batholith, Washington, determined using aluminum-in-hornblende barometry: Implications for northward translation of Baja British Columbia, *Geol. Soc. Am. Bull.*, 108, 471–488.

Ali, J.R., Ward, D.J., King, C. & Abrajevitch, A., 2003. First Paleogene sedimentary rock palaeomagnetic pole from stable western Eurasian and tectonic implications, *Geophys. J. Int.*, 154, 463–470.

Anson, G.L. & Kodama, K.P., 1987. Compaction-induced inclination shallowing of the post-depositional remanent magnetization in a synthetic sediment, *Geophys. J. Roy. Astron. Soc.*, 88, 673–692.

Arason, P. & Levi, S., 1990. Compaction and inclination shallowing in deep-sea sediments from the Pacific Ocean, *J. Geophys. Res.*, 95, 4501–4510.

Bailey, R.J., 2009. Cyclostratigraphic reasoning and orbital time calibration, *Terra Nova*, 21, 340–351.

Banerjee, S.K., Elmore, R.D. & Engel, M.H., 1997. Chemical remagnetization and burial diagenesis; testing the hypothesis in the Pennsylvania Belden Formation, Colorado, *J. Geophys. Res.*, 102, 24, 825–824, 842.

Barton, C.E. & McElhinny, M.W., 1979. Detrital remanent magnetisation in five slowly redeposited long cores of sediment, *Geophys. Res. Lettrs*, 6, 229–232.

Barton, C.E., McElhinny, M.W. & Edwards, D.J., 1980. Laboratory studies of depositional DRM, *Geophys. J. Roy. Astron. Soc.*, 61, 355–377.

Bennett, R.H., Bryant, W.R. & Keller, G.H., 1981. Clay fabric of selected submarine sediments: Fundamental properties and models, *J. Sediment. Petrol.*, 51, 217–232.

Berner, R.A., 1981. A new geochemical classification of sedimentary environments, *J. Sediment. Petrol.*, 51, 359–365.

Besse, J. & Courtillot, V., 2002. Apparent and true polar wander and the geometry of the geomagnetic field over the last 200 Myr., *J. Geophys. Res.*, 107, doi: 10.1029/2000JB000050.

Bilardello, D. & Kodama, K.P., 2009a. Measuring remanence anisotropy of hematite in red beds: anisotropy of high-field isothermal remanence magnetization (hf-AIR), *Geophys. J. Int.*, 178, 1260–1272.

Bilardello, D. & Kodama, K.P., 2009b. Palaeomagnetism and magnetic anisotropy of Carboniferous red beds from the Maritime Provinces of Canada: evidence for shallow palaeomagnetic inclnations and implications for North American apparent polar wander, *Geophys. J. Int.*, 180, 1013–1029.

Bilardello, D. & Kodama, K.P., 2010a. A new inclination shallowing correction of the Mauch Chunk Formation of Pennsylvania, based on high-field AIR results: Implications for the Carboniferous North American APW path and Pangea reconstructions, *Earth Planet. Sci. Lettrs*, 299, 218–227.

Bilardello, D. & Kodama, K.P., 2010b. Rock magnetic evidence for inclination shallowing in the early Carboniferous Deer Lake Group red beds of western Newfoundland, *Geophys. J. Int.*, 181, 275–289.

Bilardello, D. Jezek, J., & Kodama, K.P., 2011. Propagating and incorporating error in anisotropy-based inclination corrections, *Geophys. J. Int.*, 187, 75–84.

Bingham, C., 1974. An antipodally symmetric distribution on a sphere, *Ann. Statist.*, 2, 1201–1225.

Blakemore, R.P., 1975. Magnetotactic bacteria, *Science*, 190, 377–379.

Bloemendal, J. & de Menocal, P., 1989. Evidence for a change in the periodicitiy of tropical climate cycles at 2.4 Myr from whole-core magnetic susceptibility measurements, *Nature*, 342, 897–900.

Bloemendal, J., Lamb, B. & King, J.W., 1988. Paleoenvironmental implications of rock-magnetic properties of late Quaternary sediment cores from the eastern Equatorial Atlantic, *Paleoceanography*, 3, 61–87.

Paleomagnetism of Sedimentary Rocks: Process and Interpretation, First Edition. Kenneth P. Kodama.
© 2012 Kenneth P. Kodama. Published 2012 by Blackwell Publishing Ltd.

Bloemendal, J., King, J.W., Hall, F.R. & Doh, S.-J., 1992. Rock magnetism of Late Neogene and Pleistocene Deep Sea sediments: Relationship to sediment source, diagenetic processes and sediment lithology, *J. Geophys. Res.*, 97, 4361–4375.

Blow, R.A. & Hamilton, N., 1978. Effect of compaction on the acquisition of a detrital remanent magnetism in fine-grained sediments, *Geophys. J. Roy. Astron. Soc.*, 52, 13–23.

Blumstein, A.M., Elmore, R.D., Engel, M.H., Eliot, C. & Basu, A., 2004. Paleomagnetic dating of burial diagenesis in Mississippian carbonates, Utah, *J. Geophys. Res.*, 109(B4), B04101, doi:10.1029/2003JB002698.

Bohacs, K.M., Carroll, A.R., Neal, J.E. & Mankiewicz, P.J., 2000. Lake-basin type, source potential, and hydrocarbon character: an integrated sequence-stratigraphic-geochemical framework. In *Lake Basins through Space and Time*, Gierlowski-Kordesch, E.H. & Kelts, K.R. (eds), AAPG, Tulsa, OK, Studies in Geology, pp. 3–34.

Bond, G.C., Kromer, B., Beer, J., Muscheler, R., Evans, M.N., Showers, W., Hoffmann, S., Lotti-Bond, R., Hajdas, I. & Bonani, G., 2001. Persistent solar influence on North Atlantic climate during the Holocene, *Science*, 294, 2130–2136.

Borradaile, G.J., 1981. Particulate flow of rock and the formation of rock cleavage, *Tectonophysics*, 72, 305–321.

Borradaile, G.J., 1993. Strain and magnetic remanence, *J. Struct. Geol.*, 15, 383–390.

Borradaile, G.J., 1994. Remagnetization of a rock analogue during experimental triaxial deformation, *Phys. Earth Planet. Int.*, 83, 147–163.

Borradaile, G.J., 1996. Experimental stress remagnetization of magnetite, *Tectonophysics*, 261, 229–248.

Borradaile, G.J., 1997. Deformation and paleomagnetism, *Surv. Geophys.*, 18, 405–435.

Borradaile, G.J & Jackson, M., 2004. Anisotropy of magnetic susceptibility (AMS); magnetic petrofabrics of deformed rocks, *Geol. Soc. Spec. Publ.*, 238, 299–360.

Bossart, P., Ottiger, R. & Heller, F., 1990. Rock magnetic properties and structural development in the core of the Hazara-Kashmir Syntaxis, NE Pakistan, *Tectonics*, 9, 103–121.

Brachfeld, S.A. & Banerjee, S.K., 2000. A new high-resolution geomagnetic relative paleointensity record for the North American Holocene; a comparison of sedimentary and absolute intensity data, *J. Geophys. Res.*, 105, 821–834.

Bressler, S.L. & Elston, D.P., 1980. Declination and inclination errors in experimentally deposited specularite-bearing sand, *J. Geophys. Res.*, 85, 339–355.

Buchner, K. & Grapes, R., 2011. *Petrogenesis of Metamorphic Rocks*, 7th edition, Springer-Verlag, Heidelberg, New York, pp. 21–56.

Butler, R.F., 1992. *Paleomagnetism: Magnetic Domains to Geologic Terranes*, Blackwell Publishing, Boston.

Butler, R.F. & Banerjee, S.K., 1975. Theoretical single-domain grain size range in magnetite and titanomagnetite, *J. Geophys. Res.*, 80, 4049–4058.

Butler, R.F. & Taylor, L.H., 1978. A middle Paleocene paleomagnetic pole from the Nacimiento Formation, San Juan Basin, New Mexico, *Geology*, 6, 495–498.

Butler, R.F., Dickinson, W.R. & Gehrels, G.E., 1991. Paleomagnetism of coastal California and Baja California: Alternatives to large-scale northward transport, *Tectonics*, 10, 561–576.

Butler, R.F., Gehrels, G.E., Crawford, M.L. & Crawford, W.A., 2001. Paleomagnetism of the Quottoon plutonic complex in the Coast Mountains of British Columbia and Southeastern, Alaska; evidence for tilting during uplift, *Can. Jour. Earth Sci.*, 38, 1367–1385.

Cairanne, B., Aubourg, C., Pozzi, J.-P., Moreau, M.-G., Decamps, T. & Marolleau, G., 2004. Laboratory chemical remanent magnetization in a natural claystone; a record of two magnetic polarities, *Geophys. J. Int.*, 159, 906–916.

Canfield, D.E. & Berner, R.A., 1987. Dissolution and pyritization of magnetite in anoxic marine sediments, *Geochim. Cosmochim. Acta*, 51, 645–659.

Carey, S.W., 1955. The orocline concept in geotectonics, *Proc. Roy. Soc. Tasmania*, 89, 255–288.

Carter-Stiglitz, B., Moskowitz, B.M. & Jackson, M., 2001. Unmixing magnetic assemblages and the magnetic behavior of biomodal mixtures, *J. Geophys. Res.*, 106, 26397–26412.

Carter-Stiglitz, B., Valet, J.-P. & LeGoff, M., 2006. Constraints on the acquisition of remanent magnetization in fine-grained sediments imposed by redeposition experiments, *Earth Planet. Sci. Lettrs*, 245, 427–437.

Cawthorn, R.G., 1996. *Layered Intrusions*, Elsevier Science, Developments in Petrology 15, Amsterdam.

Cederquist, D.P., Van der Voo, R. & van der Pluijm, B., 2006. Syn-folding remagnetization of Cambro-Ordovician carbonates from the Pennsylvania Salient post-dates oroclinal rotation, *Tectonophysics*, 422, 41–54.

Celaya, M.A. & Clement, B.M., 1988. Inclination shallowing in deep-sea sediments from the North Atlantic, *Geophys. Res. Lettrs*, 15, 52–55.

Champion, D.E., Howell, D.G. & Gromme, C.S., 1984. Paleomagnetic and geologic data indicating 2500 km of northward displacement for the Salinian and related terranes, California, *J. Geophys. Res.*, 89, 7736–7752.

Chang, S.-B. & Kirschvink, J.L., 1989. Magnetofossils, the magnetization of sediments and the evolution of magnetite biomineralization, *Ann. Rev. of Earth Planet. Sci.*, 17, 169–195.

Channell, J.E.T., Curtis, J.H. & Flower, B.P., 2004. The Matuyama-Brunhes boundary interval (500–900 ka) in North Atlantic drift sediments, *Geophys. J. Int.*, 158, 489–505.

Channell, J.E.T., Casellato, C.E., Muttoni, G. & Erba, E., 2010. Magnetostratigraphy, nannofossil stratigraphy and apparent polar wander for Adria-Africa in the Jurassic–Cretaceous boundary interval, *Paleogeog., Paleoclimat. Paleoecol.*, 293, 51–75.

Chen, A.P., Egli, R. & Moskowitz, B.M., 2007. First-order reversal curve (FORC) diagrams of natural and cultured biogenic magnetic particles, *J. Geophys. Res.*, 112, B08S90.

Cioppa, M.T. & Kodama, K.P., 2003a. Environmental magnetic and magnetic fabric studies in Lake Waynewood, northeastern Pennsylvania, USA: Evidence for changes in watershed dynamics, *J. Paleolimnology*, 29, 61–78.

Cioppa, M.T. & Kodama, K.P., 2003b. Evaluation of paleomagnetic and finite strain relationships due to the Alleghanian Orogeny in the Mississippian Mauch Chunk Formation, Pennsylvania, *J. Geophys. Res.*, 108, EPM8 1–16.

Cisowski, S., 1981. Interacting vs. non-interacting single domain behavior in natural and synthetic samples, *Phys Earth Planet. Int.*, 26, 56–62.

Clement, B., 1991. Geographical distribution of transitional VGPs: Evidence for non-zonal equatorial symmetry during the Matuyama-Brunhes geomagnetic reversal, *Earth Planet. Sci. Lettrs*, 104, 48–58.

Cogne, J.P., 1987. Experimental and numerical modeling of IRM rotation in deformed synthetic samples, *Earth Planet. Sci. Lettrs*, 86, 39–45.

Cogne, J.P. & Perroud, H., 1985. Strain removal applied to paleomagnetic directions in an orogenic belt; the Permian red slates of the Alpes Maritimes, France, *Earth Planet. Sci. Lettrs*, 72, 125–140.

Collinson, D.W., 1965. DRM in sediments, *J. Geophys. Res.*, 70, 4663–4668.

Collombat, H., Rochette, P. & Kent, D.V., 1993. Detection and correction of inclination shallowing in deep-sea sediments using the anisotropy of anhysteretc remanence, *Bull. Soc. Geol. France*, 164, 103–111.

Constable, C. & Parker, R.L., 1988. Statistics of the geomagnetic secular variation for the past 5 m.y., *J. Geophys. Res.*, 93, 11569–11581.

Constable, C. & Johnson, C.L., 1999. Anisotropic paleosecular variation models: Implications for geomagnetic observables, *Phys. Earth Planet. Inter.*, 115, 35–51.

Cowan, D.S., Brandon, M.T. & Garver, I.J., 1997. Geologic tests of hypotheses for large coastwise displacements: A critique illustrated by the Baja British Columbia controversy, *Am. J. Sci.*, 297, 117–173.

Cox, E., Elmore, R.D. & Evans, M.E., 2005. Paleomagnetism of Devonian red beds in the Appalachian Plateau and Valley and Ridge Provinces, *J. Geophys. Res.*, 110(B8), doi: 10.1029/2005JB003640.

Creer, K.M. & Tucholka, P., 1982. Secular variation as recorded in lake sediments; a discussion of North American and European results, *Phil. Trans. Roy Soc. London, Series A*, 306, 87–102.

Day, R., Fuller, M.D. & Schmidt, P.W., 1977. Hysteresis properties of titanomagnetites: grain size and composition dependence, *Phys Earth Planet. Int.*, 13, 260–266.

de Menocal, P.B., Ruddiman, W.F. & Kent, D.V., 1990. Depth of post-depositional remanence acquisition in deep-sea sediments: a case study of the Brunhes-Matuyama reversal and oxygen isotope Stage 19.1, *Earth Planet. Sci. Lettrs*, 99, 1–13.

Deamer, G.A. & Kodama, K.P., 1990. Compaction-induced inclination shallowing in synthetic and natural clay-rich sediments, *J. Geophys. Res.*, 95, 4511–4529.

Dekkers, M.J. & Linssen, J.H., 1991. Grain size separation of hematite in the <5 micron range for rock magnetic investigations, *Geophys. J. Int.*, 104, 423–427.

Denham, C.R. & Chave, A.D., 1982. Detrital remanent magnetization; viscosity of the lock-in zone, *J. Geophys. Res.*, 87, 7126–7130.

Diehl, J.F., Beck, M.E., Beske-Diehl, S., Jacobson, D. & Hearn, B.C. Jr., 1983. Paleomagnetism of Late Cretaceous-early Tertiary north-central Montana alkalic province, *J. Geophys. Res.*, 88, 10593–10609.

Donovan, T.J., Forgey, R.L. & Roberts, A.A., 1979. Aeromagnetic detection of diagenetic magnetite over oil fields, *Am. Assoc. Pet. Geol. Bull.*, 63, 245–248.

Dubiel, R.F. & Smoot, J.P., 1994. Criteria for intepreting paleoclimate from red beds; a tool for Pangean reconstructions, *Memoir-Canadian Soc. Petrol. Geol.*, 17, 295–310.

Dunlop, D.J. & Ozdemir, O., 1997. *Rock Magnetism: Fundamentals and Frontiers*, Cambridge University Press, Cambridge.

Dupont-Nivet, G., Guo, Z., Butler, R.F. & Jia, C., 2002. Discordant paleomagnetic direciton in Miocene rocks from the central Tarim Basin; evidence for local deformation and inclination shallowing, *Earth Planet. Sci. Lettrs*, 199, 473–482.

Dupont-Nivet, G., Lippert, P.C., van Hinsbergen, D.J.J., Meijers, M.J.M. & Kapp, P., 2010. Palaeolatitude and age of the Indo-Asia collision; palaeomagnetic constraints, *Geophys. J. Int.*, 182, 1189–1198.

Egli, R., 2004. Characterization of individual rock magnetic components by analysis of remanence cruves, 1. Unmixing natural sediments, *Stud. Geophys. Geod.*, 48, 391–446.

Egli, R., Chen, A.P., Winklhofer, M., Kodama, K.P. & Horng, C.-S., 2010. Detection of noninteracting single domain particles using first-order reversal curve diagrams, *Geochem. Geophys. Geosyst.*, 11, Q01Z11.

Ellwood, B.B., 1984. Bioturbation; some effects on remanent magnetization acquisition, *Geophys. Res. Lettrs*, 11, 653–655.

Ellwood, B.B., Kafafy, A.M., Kassab, A., Abdeldayem, Obaldalla, N., Howe, R.W. & Sikora, P., 2010. Magnetostratigraphy susceptiblity used for high-resolution correlation among Santonian (Upper Cretaceous) marine sedimentary sequences in the U.S. Western Interior Seaway and the western Sinai Peninsula, Egypt. In *Application of Modern*

Stratigraphic Techniques: Theory and Case Histories, Radcliffe, K.T. & Zaitlin, B.A. (eds), SEPM, Houston, pp. 155–166.

Ellwood, B.B., Algeo, T.J., El Hassani, A., Tomkin, J.H. & Rowe, H., 2011. Defining the timing and duration of the Kacak Interval within the Eifelian/Givetian boundary GSSP, Mech Irdane, Morocco using geochemical and magnetic susceptibility patterns, *Paleogeog., Paleoclimat. Paleoecol.*, 304, 74–84.

Elmore, R.D., London, D., Bagley, D., Fruit, D. & Gao, G., 1993. Remagnetization by basinal fluids; testing the hypothesis in the Viola Limestone, southern Oklahoma, *J. Geophys. Res.*, 98, 6237–6254.

Elmore, R.D., Foucher, J.L.-E., Evans, M.E., Lewchuk, M.T. & Cox, E., 2006. Remagnetization of the Tonoloway Formation and the Helderberg Group in the Central Appalachians: testing the origin of syntilting magnetizations, *Geophys. J. Int.*, 166, 1062–1076.

Enkin, R.J., 2003. The direction-correction tilt test; an all-purpose tilt/fold test for paleomagnetic studies, *Earth Planet. Sci. Lettrs*, 212, 151–166.

Enkin, R.J., Baker, J. & Mustard, P.S., 2001. Paleomagnetism of the Upper Cretaceous Nanaimo Group, southwestern Canadian Cordillera, *Can. Jour. Earth Sci.*, 38, 1403–1422.

Evans, M.E. & Maillol, J.M., 1986. A palaeomagnetic investigation of a Permian redbed sequence from a mining drill core, *Geophys. J. Roy. Astron. Soc.*, 87, 411–419.

Evans, M.E. & Heller, F., 2003. *Environmental Magnetism: Principles and Applications of Environmagnetics*, Academic Press, San Diego.

Facer, R., 1983. Foding strain and Graham's fold test in paleomagnetic investigations, *Geophys. J. Roy. Astron. Soc.*, 72, 165–171.

Fisher, R.A., 1953. Dispersion on a sphere, *Proc. Roy. Soc. London, Ser. A*, 217, 295–305.

Francis, P., 1993. *Volcanoes: A Planetary Perspective*, Oxford University Press, Oxford.

Franke, C., von Dobeneck, T., Drury, M.T., Meeldijk, J.D. & Dekkers, M.J., 2007. Magnetic petrology of equatorial Atlantic sediments: Electron microscopy results and their implications for environmental magnetic interpretation, *Paleoceanography*, 22, PA4207, doi:10.1029/2007PA001442.

Froelich, P.N., Klinkhammer, G.P., Bender, M.L., Luedtke, N.A., Heath, G.R., Cullen, D., Dauphin, P., Hammond, D., Hartman, B. & Maynard, V., 1979. Early oxidation of organic matter in pelagic sediments of the eastern equatorial Atlantic: suboxic diagenesis, *Geochim. Cosmochim. Acta*, 43, 1075–1090.

Fry, N., 1979. Density distribution techniques and strained length methods for determination of finite strains, *J. Struct. Geol.*, 1, 221–229.

Garces, M., Pares, J.M. & Cabrera, L., 1996. Inclination error linked to sedimentary facies in Miocene detrital sequences from the Valles-Penedes Basin (NE Spain). In *Palaeomagnetism and Tectonics of the Mediterranean Region*, Morris, A. & Tarling, D.H. (eds), Geological Society, London, Special Publication, pp. 91–99.

Geissman, J.W. & Harlan, S.S., 2002. Late Paleozoic remagnetization of Precambrian crystalline rocks along the Precambrian/Carboniferous nonconformity, Rocky Mountains, a relationship among deformation, remagnetization, and fluid migration, *Earth Planet. Sci. Lettrs*, 203, 905–924.

Gilder, S., Chen, Y. & Sen, S., 2001. Oligo-Miocene magnetostratigraphy and rock magnetism of the Xishuigou section, Subei (Gansu Province, western China) and implications for shallow inclinations in Central Asia, *J. Geophys. Res.*, 106, 30505–30521.

Gilder, S., Chen, Y., Cogne, J.P., Tan, X., Courtillot, V., Sun, D. & Li, Y., 2003. Paleomagnetism of Upper Jurassic to Lower Cretaceous volcanic and sedimentary rocks from the western Tarim Basin and implications for inclination shallowing and absolute dating of the M-0 (ISEA?) chron, *Earth Planet. Sci. Lettrs*, 206, 587–600.

Gill, J.D., Elmore, R.D. & Engel, M.H., 2002. Chemical remagnetization and clay diagenesis; testing the hypothesis in the Cretaceous sedimentary rocks of northwestern Montana, *Phys. Chem. Earth*, 27, 25–31.

Gillett, S.L. & Karlin, R., 2004. Pervasive late Paleozoic-Triassic remagnetization of miogeoclinal carbonate rocks in the Basin and Range and vicinity, SW USA; regional results and possible tectonic implications, *Phys Earth Planet. Int.*, 141, 95–120.

Goldhammer, R.K., Dunn, P.A. & Hardie, L.A., 1990. Depositional cycles, composite sea-level changes, cycle stacking patterns, and the hierarchy of stratigraphic forcing: Examples from Alpine Triassic platform carbonates., *Geol. Soc. Am. Bull.*, 102, 535–562.

Goldhammer, R.K., Dunn, P.A., Harris, M.T. & Hardie, L.A., 1991. Sequence stratigraphy and systems tract development of the Latemar Platform, Middle Triassic of the Dolomites; outcrop calibration keyed by cycle stacking patterns, *AAPG Bulletin*, 75(3), 582.

Gradstein, F. M., Ogg, J.G. & Smith, A. (eds) 2004. *A Geologic Time Scale*, Cambridge University Press, New York.

Graham, J.W., 1949. The stability and significance of magnetism in sedimentary rocks, *J. Geophys. Res.*, 54, 131–167.

Graham, S., 1974. Remanent magnetization of modern tidal flat sediments from San Francisco Bay, Calif., *Geology*, 2, 223–226.

Granar, L., 1958. Magnetic measruements on Swedish varved sediments, *Ark. Geofys.*, 3, 1–40.

Gray, M.B. & Mitra, G., 1993. Migration of deformation fronts during progressive deformation; evidence from detailed structural studies in the Pennsylvania Anthracite region, U.S.A., *J. Struct. Geol.*, 15, 435–449.

Gray, M.B. & Stamatakos, J., 1997. New model for evolution of fold and thrust belt curvature based on integrated structural and paleomagnetic results from the Pennsylvania salient, *Geology*, 25, 1067–1070.

Griffiths, D.H., King, R.F., Rees, A.I. & Wright, A.E., 1960. Remanent magnetism of some recent varved sediments, *Proc. Roy. Soc. London, Ser. A*, 256, 359–383.

Gudmundsson, A., 2009. Toughness and failure of volcanic edifices, *Tectonophysics*, 471, 27–35.

Guyodo, Y. & Valet, J.-P., 1996. Relative variations in geomagnetic intensity from sedimentary records, the past 200,000 years, *Earth Planet. Sci. Lettrs*, 143, 23–36.

Guyodo, Y. & Valet, J.-P., 1999. Global changes in intensity in the Earth's magnetic field during the past 800 kyr, *Nature*, 399, 249–252.

Hagstrum, J.T. & Champion, D.E., 1995. Late Quaternary geomagnetic secular variation from historical and ^{14}C-dated lava flows on Hawaii, *J. Geophys. Res.*, 100, 24393–24403.

Hallam, D.F. & Maher, B.A., 1994. A record of reversed polarity carried by the iron sulphide greigite in British early Pleistocene sediments, *Earth Planet. Sci. Lettrs*, 121, 71–80.

Hamano, Y., 1980. An experiment on the post-depositional remanent magnetization in artificial and natural sediments, *Earth Planet. Sci. Lettrs*, 51, 221–232.

Hankard, F., Cogne, J.P., Kravchinsky, V., Carporzen, L., Bayasgaian, A. & Lkhagvadorj, P., 2007. New Tertiary paleomagnetic poles from Mongolia nad Siberia at 40, 30, 20, and 13 Ma; clues on the inclination shallowing problem in central Asia, *J. Geophys. Res.*, 112, B02101.

Harris, S.E. & Mix, A.C., 2002. Climate and tectonic influences on continental erosion of tropical South America, 0–13 Ma, *Geology*, 30, 447–450.

Harrison, C.G.A., 1966. Paleomagnetism of deep-sea sediments, *J. Geophys. Res.*, 71, 3033–3043.

Hartl, P. & Tauxe, L., 1996. A precursor to the Matuyama/Brunhes transition-field instability as recorded in pelagic sediments, *Earth Planet. Sci. Lettrs*, 138, 121–135.

Hays, J.D., Imbrie, J. & Shackleton, N.J., 1976. Variations in the Earth's orbit; pacemaker of the ice ages, *Science*, 194, 1121–1132.

Heller, F., Shen, C.D., Beer, J., Liu, X.M., Liu, T.S., Bronger, A., Suter, M. & Bonani, G., 1993. Quantitative estimates and palaeoclimatic implications of pedogenic ferromagnetic mineral formation in Chinese loess, *Earth Planet. Sci. Lettrs*, 114, 385–390.

Hillhouse, J.W., 2010. Clockwise rotation and implications for northward drift of the western Transverse Ranges from paleomagnetism of the Piuma Member, Sespe Formation, near Malibu, California, *Geochem. Geophys. Geosyst.*, 11, Q07005.

Hinnov, L.A., 2000. New perspectives on orbitally forced stratigraphy, *Annu. Rev. Earth Planet. Sci.*, 28, 419–475.

Hinnov, L.A., Anastasio, D.J., Latta, D.K., Kodama, K.P. & Elrick, M., 2009. Milankovitch-controlled paleoclimate signal recorded by rock magnetics, Lower Cretaceous platform carbonates of northern Mexico, *AAPG Datapages/Search and Discovery*, #40388.

Hirt, A.M., Lowrie, W. & Pfiffner, O.A., 1986. A paleomagnetic study of tectonically deformed red beds of the lower Glarus nappe complex, eastern Switzerland, *Tectonics*, 5, 723–731.

Hnat, J.S., van der Pluijm, B. & Van der Voo, R., 2009. Remagnetization in the Tennessee Salient, Southern Appalachians, USA; constraints on the timing of deformation, *Tectonophysics*, 474, 709–722.

Hodych, J.P. & Bijaksana, S., 1993. Can remanence anisotropy detect paleomagnetic inclination shallowing due to compaction? A case study using Cretaceous deep-sea sediments, *J. Geophys. Res.*, 98, 22429–22441.

Hodych, J.P. & Buchan, K.L., 1994. Early Silurian palaeolatitude of the Springdale Group redbeds of central Newfoundland; a palaeomagnetic determination with a remanence anistropy test for inclination error, *Geophys. J. Int.*, 117, 640–652.

Hodych, J.P., Bijaksana, S. & Paetzold, R., 1999. Using magnetic anisotropy to correct for paleomagnetic inclination shallowing in some magnetite-bearing deep-sea turbidites and limestones, *Tectonophysics*, 307, 191–205.

Hooghiemstra, H., Melice, J.-L., Berger, A. & Shackleton, N.J., 1993. Frequency spectra and paleoclimatic variability of the high-resolution 30–1450 ka Funza I pollen record (Eastern Cordlilera, Colombia), *Quat. Sci. Rev.*, 12, 141–156.

Horng, C.-S. & Roberts, A.P., 2006. Authigenic or detrital origin of pyrrothite in sediments? Resolving a paleomagnetic conundrum, *Earth Planet. Sci. Lettrs*, 241, 750–762.

Horng, C.-S., Torii, M., Shea, K.S. & Kao, S.J., 1998. Inconsistent magnetic polarities between greigite- and pyrrhotite/magnetite-bearing marine sediments from the Tsailiao-chi section, southwestern Taiwan, *Earth Planet. Sci. Lettrs*, 164, 467–481.

Hounslow, M.W. & Maher, B.A., 1996. Quantitative extraction and analysis of carriers of magnetization in sediments, *Geophys. J. Int.*, 124, 57–74.

Hounslow, M.W. & Maher, B.A., 1999. Source of the climate signal recorded by magnetic susceptibility variations in Indian Ocean sediments, *J. Geophys. Res.*, 104, 5047–5061.

Hounslow, M.W. & Muttoni, G., 2010. The geomagnetic polarity timescale for the Triassic: linkage to stage boundary definitions, *Geol. Soc. London, Spec. Publ.*, 334, 61–102.

Housen, B.A., van der Pluijm, B. & Van der Voo, R., 1993. Magnetite dissolution and neocrystallization during cleavage formation; paleomagnetic study of the Martinsburg Formation, Lehigh Gap, Pennsylvania, *J. Geophys. Res.*, 98, 13799–13813.

Hudson, M.R., Reynolds, R.L. & Fishman, N.S., 1989. Synfolding magnetization in the Jurassic Preuss Sandstone, Wyoming-Idaho-Utah thrust belt, *J. Geophys. Res.*, 94, 13681–13705.

Huesing, S.K., Dekkers, M.J., Franke, C. & Krijgsman, W., 2009. The Tortonian reference section at Monte dei Corvi

(Italy); evidence for early remanence acquistion in greigite-bearing sediments, *Geophys. J. Int.*, 179, 125–143.

Iosifidi, A.G., MacNiocaill, C., Khramov, A.N., Dekkers, M.J. & Popov, V.V., 2010. Palaeogeographic implications of differential inclination shallowing in Permo-Carboniferous sediments from the Donets Basin, Ukraine, *Tectonophysics*, 490, 229–240.

Irurzun, M.A., Gogorza, C.S.G., Chaparro, M.A.E., Linio, J.M., Nunez, H., Vilas, J.F. & Sinito, A.M., 2006. Paleosecular variations recorded by Holocene-Pleistocene sediments from Lake El Trebol (Patagonia, Argentina), *Phys. Earth Planet. Int.*, 154, 1–17.

Irving, E., 1957. Origin of the paleomagnetism of the Torridonian sandstones of north-west Scotland, *Phil. Trans. Roy. Soc. London, Ser. A*, 250, 100–110.

Irving, E. & Major, A., 1964. Post-depositional detrital remanent magnetization in a synthetic sediment, *Sedimentology*, 3, 135–143.

Iwaki, H. & Hayashida, A., 2003. Paleomagnetism of Pleistocene widespread tephra deposits and its implication for tectonic rotation in central Japan, *Island Arc*, 12, 46–60.

Jackson, M., 1990. Diagenetic sources of stable remanence in remagnetized Paleozoic cratonic carbonates: A rock magnetic study, *J. Geophys. Res.*, 95, 2753–2761.

Jackson, M., Banerjee, S.K., Marvin, J.A., Lu, R. & Gruber, W., 1991. Detrital remanence, inclination errors, and anhysteretic remanence anisotropy: Quantitative model and experimental results, *Geophys. J. Int.*, 104, 95–103.

Jackson, M., Sun, W.-W. & Craddock, J.P., 1992. The rock magnetic fingerprint of chemical remagnetization in midcontinental Paleozoic carbonates, *Geophys. Res. Lettrs*, 19, 781–784.

Jeffery, G.B., 1923. On the motion of ellipsoidal particles immersed in a viscous fluid, *Proc. Roy. Soc. London, Ser. A*, 102, 161–179.

Jerolmack, D.J. & Paola, C., 2010. Shredding of enviornmental signals by sediment transport, *Geophys. Res. Lettrs*, 37, L19401.

Jiang, W., Horng, C.-S., Roberts, A.P. & Peacor, D.R., 2001a. Contradictory magnetic polarities in sediments and variable timing of neoformation of authigenic greigite, *Earth Planet. Sci. Lettrs*, 193, 1–12.

Jiang, W.T., Horng, C.-S., Roberts, A.P. & Peacor, D.R., 2001b. Contradictory magnetic polarities in sediments and variable timing of neoformation of authigenic greigite, *Earth Planet. Sci. Lettrs*, 193, 1–12.

Johnson, E.A., Murphy, T. & Torreson, O.W., 1948. Pre-history of the Earth's magnetic field, *Terr. Magn. Atmos Elect.*, 53, 349–372.

Jovane, L., Acton, G., Florindo, F. & Verosub, K.L., 2008. Geomagnetic field behavior at high latitudes from a paleomagnetic record from Eltanin core 27–21 in the Ross Sea sector, Antarctica, *Earth Planet. Sci. Lettrs*, 267, 435–443.

Karlin, R., 1990. Magnetite diagenesis in marine sediments from the Oregon continental margin, *J. Geophys. Res.*, 95, 4405–4419.

Karlin, R. & Levi, S., 1983. Diagenesis of magnetic minerals in Recent haemipelagic sediments, *Nature*, 303, 327–330.

Karlin, R. & Levi, S., 1985. Geochemical and sedimentological control of the magnetic properties of hemipelagic sediments, *J. Geophys. Res.*, 90, 10373–10392.

Kasten, S., Zabel, M., Heuer, V. & Hensen, C., 2003. Processes and signals of nonsteady-state diagenesis in deep-sea sediments and their pore waters. In *The South Atlantic in the Late Quaternary: Reconstruction of Material Budgets and Current Systems*, Wefer, G., Mulitza, S. & Ratmeyer, V. (eds), Springer-Verlag, Berlin, pp. 431–459.

Katari, K., Tauxe, L. & King, J., 2000. A reassessment of post-depositional remanent magnetism: preliminary experiments with natural sediments, *Earth Planet. Sci. Lettrs*, 183, 147–160.

Katz, B., Elmore, R.D., Cogoini, M. & Ferry, S., 1998. Widespread chemical remagnetization; orogenic fluids or burial diagenesis of clays? *Geology*, 26, 603–606.

Katz, B., Elmore, R.D., Cogoini, M., Engel, M.H. & Ferry, S., 2000. Associations between burial diagenesis of smectite, chemical remagnetization, and magnetite authigenesis in the Vocontian Trough, SE France, *J. Geophys. Res.*, 105, 851–868.

Kauahikaua, J.P., Trusdell, F. & Heliker, C., 1998. The probability of lava inundation at the proposed and existing Kulani Prison sites, *USGS Open-File Report*, 98–794.

Kawamura, K., Oda, H., Ikehara, K., Yamazaki, T., Shioi, J., Taga, S. & Hatakeyama, S., 2007. Diagenetic effect on magnetic properties of marine core sediments from the southern Okhotsk Sea, *Earth, Planets, Space*, 59, 83–93.

Keen, M.J., 1963. The magnetization of sediment cores from the eastern basin of the North Atlantic Ocean, *Deep Sea Res.*, 10, 607–622.

Kent, D.V., 1973. Post-depositional remanent magnetization in a deep-sea sediment, *Nature*, 246, 32–34.

Kent, D.V., 1979. Paleomagnetism of the Devonian Onondaga Limestone revisited, *J. Geophys. Res.*, 84, 3576–3588.

Kent, D.V., 1985. Thermoviscous remagnetization in some Appalachian limestones, *Geophys. Res. Lettrs*, 12, 805–808.

Kent, D.V., 1988. Further paleomagnetic evidence for oroclinal rotation in the central folded Appalachians from the Bloomsburg and the Mauch Chunk formations, *Tectonics*, 7, 749–759.

Kent, D.V. & Opdyke, N.D., 1985. Multicomponent magnetization from the Mississippian Mauch Chunk formation of the central Appalachians and their tectonic implications, *J. Geophys. Res.*, 90, 5371–5383.

Kent, D.V. & Schneider, D.A., 1995. Correlation of paleointensity variation records in the Brunhes/Matuyama transition interval, *Earth Planet. Sci. Lettrs*, 129, 135–144.

Kent, D.V. & Olsen, P.E., 1997. Paleomagnetism of Upper Triassic continental sedimentary rocks from the Dan River-Danville rift basin (eastern North America), *Geol. Soc. Am. Bull.*, 109, 366–377.

Kent, D.V. & Smethurst, M.A., 1998. Shallow bias of paleomagnetic inclinations in the Paleozoic and Precambrian, *Earth Planet. Sci. Lettrs*, 160, 391–402.

Kent, D.V., Muttoni, G. & Brack, P., 2004. Magnetostratigraphic confirmation of a much faster tempo for sea-level change for the Middle Triassic Latemar platform carbonates, *Earth Planet. Sci. Lettrs*, 228, 369–377.

Kent, J.T., 1982. The Fisher-Bingham distribution on the sphere, *J. R. Statist. Soc. B*, 44, 71–80.

Kilgore, B. & Elmore, R.D., 1989. A study of the relationship between hydrocarbon migration and the precipitation of authigenic magnetic minerals in the Triassic Chugwater Formation, southern Montana, *Geol. Soc. Am. Bull.*, 101, 1280–1288.

Kim, B.-Y. & Kodama, K.P., 2004. A compaction correction for the paleomagnetism of the Nanaimo Group sedimentary rocks: Implications for the Baja British Columbia hypothesis, *J. Geophys. Res.*, 109(17), doi:10.1029/2003JB002696.

Kim, B.-Y., Kodama, K.P. & Moeller, R.E., 2005. Bacterial magnetite produced in water column dominates lake sediment mineral magnetism: Lake Ely, USA, *Geophys. J. Int.*, 163, 26–37.

King, R.F., 1955. Remanent magnetism of artificially deposited sediments, *Monthly Notices Roy. Astron. Soc. Geophys. Suppl.*, 7, 115–134.

Kirker, A. & McClelland, E., 1997. Deflection of magnetic remanence during progressive cleavage development in the Pembrokeshire Old Red Sandstone, *Geophys. J. Int.*, 130, 240–250.

Kirschvink, J.L., Jones, D.S., MacFadden, B.J. & Stehli, F.G. (eds) 1985. *Magnetite Biomineralization and Magnetoreception in Organisms: A New Biomagnetism*, Plenum Publishing, New York, NY.

Kligfield, R., Lowrie, W., Hirt, A.M. & Siddans, A., 1983. Effect of progressive deformation on remanent magnetization of Permian redbeds from the Alpes Maritimes (France), *Tectonophysics*, 97, 59–85.

Knudsen, M.F., Henderson, G.M., Frank, M., MacNiocaill, C. & Kubik, P.W., 2008. In-phase anomalies in Beryllium-10 production and palaeomagnetic field behaviour during the Iceland Basin geomagnetic excursion, *Earth Planet. Sci. Lettrs*, 265, 588–599.

Kobayashi, A., Kirschvink, J.L., Nash, C.Z., Kopp, R.E., Sauer, D.A., Bertani, L.E., Voorhout, W.F. & Taguchi, T., 2006. Experimental observation of magnetosome chain collapse in magnetotactic bacteria; sedimentological, paleoamgnetic, and evolutionary implications, *Earth Planet. Sci. Lettrs*, 245, 538–550.

Kodama, K.P., 1982. Magnetic effects of maghemitization of Plio-Pleistocene marine sediments, northern California, *J. Geophys. Res.*, 87, 7113–7125.

Kodama, K.P., 1988. Remanence rotation due to rock strain during folding and stepwise application of the fold test, *J. Geophys. Res.*, 93, 3357–3371.

Kodama, K.P., 1997. A successful rock magnetic technique for correcting paleomagnetic inclination shallowing: Case study of the Nacimiento Formation, New Mexico, *J. Geophys. Res.*, 102, 5193–5205.

Kodama, K.P., 2009. Simplification of the anisotropy-based inclination correction technique for magnetite- and haematite-bearing rocks: a case study for the Carboniferous Glenshaw and Mauch Chunk Formations, North America, *Geophys. J. Int.*, 176, 467–477.

Kodama, K.P. & Goldstein, A.G., 1991. Experimental simple shear deformation of magnetic remanence, *Earth Planet. Sci. Lettrs*, 104, 80–88.

Kodama, K.P. & Davi, J.M., 1995. A compaction correction for the paleomagnetism of the Cretaceous Pigeon Point Formation of California, *Tectonics*, 14, 1153–1164.

Kodama, K.P. & Ward, P., 2001. Compaction-corrected paleomagnetic paleolatitudes for Late Cretaceous rudists along the Cretaceous California margin; evidence for less than 1500 km of post-Late Creatceous offset for Baja British Columbia, *Geol. Soc. Am. Bull.*, 113, 1171–1178.

Kodama, K.P. & Dekkers, M.J., 2004. Magnetic anisotropy as an aid to identifying CRM and DRM in red sedimentary rocks, *Stud. Geophys. Geod.*, 48, 747–766.

Kodama, K.P. & Hinnov, L.A., 2007. Mineral magneic parameters provide new evidence on the climate-driver of the Triassic Latemar carbonate cycles. in *GSA Annual Meeting*, Geol. Soc. America, Denver.

Kodama, K.P. & Hillhouse, J.W., 2011. Rock magnetic cyclostratigraphy and magnetostratigraphy of the Rainstorm Member of the Neoproterozoic Johnnie Formation indicate a 2.5 Myr duration for the negative ^{13}C isotopic anomaly. Presented at 2011 Fall Meeting, AGU, San Francisco, California, 5–9 December, abstract no. GP53A-07.

Kodama, K.P., Fahringer, P. & Moeller, R.E., 1997a. Mineral magnetic record from Lake Ely sediments, PA: A rainfall-magnetic mineral concentration correlation, *Abstr with Programs, 1997 GSA Annual Meeting*, 29, A-374.

Kodama, K.P., Lyons, J.C., Siver, P.A. & Lott, A.-M., 1997b. A mineral magnetic and scaled-chrysophyte paleolimnological study of two northwestern Pennsylvania lakes: records of fly-ash deposition, land-use change, and paleorainfall variation, *J. Paleolimnology*, 17, 173–189.

Kodama, K.P., Krakoff, M., Moeller, R.E. & Fahringer, P., 1998. New magnetic mineral results from Lake Ely, PA support rainfall-magnetic mineral concentration correlation, *Eos, Trans. AGU*, 79, S67.

Kodama, K.P., Anastasio, D.J. & Hinnov, L.A., 2009. High-resolution rock magnetic cyclostratigraphy in an Eocene flysch, Spanish Pyrenees, *Eos Trans. AGU*, 90, Abs. GP 22A-06.

Kodama, K.P., Anastasio, D.J., Newton, M.L., Pares, J.M. & Hinnov, L.A., 2010. High-resolution rock magnetic cyclostratigraphy in an Eocene flysch, Spanish Pyrenees, *Geochem. Geophys. Geosyst.*, 11, Q0AA07.

Kopp, R.E., Nash, C.Z., Kobayashi, A., Weiss, B.P., Bazylinski, D.A. & Kirschvink, J.L., 2006. Ferromagnetic resonance spectroscopy for assessment of magnetic anisotropy and magnetostatic interactions: A case study of mutant magnetotactic bacteria, *J. Geophys. Res.*, 111, doi: 10.1029/2006JB004529.

Kopp, R.E., Raub, T.D., Schumann, D., Hojatollah, V., Smirnov, A.V. & Kirschvink, J.L., 2007. Magnetofossil spike during the Paleocene–Eocene Thermal Maximum: Ferromagnetic resonance, rock magnetic, and electronic microscopy evidence from Ancora, New Jersey, USA, *Paleoceanography*, 22, PA4103.

Krijgsman, W. & Tauxe, L., 2004. Shallow bias in Mediterranean paleomagnetic directions caused by inclination error, *Earth Planet. Sci. Lettrs*, 222, 685–695.

Krijgsman, W. & Tauxe, L., 2006. E/I corrected paleolatitudes for the sedimentary rocks of the Baja British Columbia hypothesis, *Earth Planet. Sci. Lettrs*, 242, 205–216.

Kruiver, P., Dekkers, M.J. & Langereis, C.G., 2000. Secular variation in Permian red beds from Dome de Barrot, SE France, *Earth Planet. Sci. Lettrs*, 179, 205–217.

Kruiver, P.P., Dekkers, M.J. & Heslop, D., 2001. Quantification of magnetic coercivity components by the analysis of acquisition curves of isothermal remanent magnetization, *Earth Planet. Sci. Lettrs*, 189, 269–276.

Kukla, G., An, Z.S., Melice, J.-L., Gavin, J. & Xiao, J.L., 1990. Magnetic susceptibility record of Chinese Loess, *Trans. R. Soc. Edinb.: Earth Sci.*, 81, 263–288.

Kumar, A.A., Purnachandra Rao, V., Patil, S.K., Kessarkar, P. & Thamban, M., 2005. Rock magnetic records of the sediments of the eastern Arabian Sea; evidence for late Quaternary climatic change, *Marine Geology*, 220, 59–82.

Larrasoana, J.C., Pares, J.M. & Pueyo, E.L., 2003. Stable Eocene magnetizion carried by magnetite and iron sulphides in marine marls (Pamplona-Arguis Formation, southern Pyrenees, northern Spain), *Stud. Geophys. Geod.*, 47, 237–254.

Larson, E.E., Walker, T.R., Patterson, P.E., Hoblitt, R.P. & Rosenbaum, J.G., 1982. Paleomagnetism of the Moenkopi Formation, Colorado Plateau; basis for long-term model of acquisition of chemical remanent magnetism in red beds, *J. Geophys. Res.*, 87, 1081–1106.

Laskar, J., Robutel, P., Joutel, F., Gastineau, M., Correia, A.C.M. & Levrard, B., 2004. A long-term numerical solution for the insolation quantities of the Earth, *Astron. and Astrophys.*, 428, 261–285.

Laskar, J., Fienga, A., Gastineau, M. & Manche, H., 2011. La2010: A new orbital solution for the long term motion of the Earth, *Earth and Planet. Astrophys.*, arXiv:1103.1084v1.

Latta, D.K., 2005. Structural, lithotectonic, and rock magnetic studies of decollement folding, Coahuila marginal folded province, northeast Mexico, PhD thesis, Lehigh University, Bethlehem, PA.

Latta, D.K., Anastasio, D.J., Hinnov, L.A., Elrick, M. & Kodama, K.P., 2006. Magnetic record of Milankovitch rhythms in lithological noncyclic marine carbonates, *Geology*, 34, 29–32.

Leslie, B.W., Lund, S.P. & Hammond, D., 1990. Rock magnetic evidence for the dissolution and authigenic growth of magnetic minerals witin anoxic marine sediments of the California continental borderland, *J. Geophys. Res.*, 95, 4437–4452.

Levi, S. & Banerjee, S.K., 1990. On the origin of inclination shallowing in redeposited sediments, *J. Geophys. Res.*, 95, 4383–4389.

Lewchuk, M.T., Evans, M.E. & Elmore, R.D., 2003. Synfolding remagnetization and deformation: results from Palaeozoic sedimentary rocks in West Virginia, *Geophys. J. Int.*, 152, 266–279.

Li, Y.-X., Yu, Z., Kodama, K.P. & Moeller, R.E., 2006. A 14,000 year environmental change history revealed by mineral magnetic data from White Lake, New Jersey, USA, *Earth Planet. Sci. Lettrs*, 246, 27–40.

Li, Y.-X., Yu, Z. & Kodama, K.P., 2007. Sensitive moisture response to Holocene millenial-scale climate variations in the Mid-Atlantic region, USA, *The Holocene*, 17, 3–8.

Liddicoat, J.C. & Coe, R.S., 1979. Mono Lake geomagnetic excursion, *J. Geophys. Res.*, 84, 261–271.

Liebes, E. & Shive, P.N., 1982. Magnetization acquistion in two Mesozoic red sandstones, *Phys Earth Planet. Int.*, 30, 396–404.

Liu, J., Zhu, R., Roberts, A.P., Li, Q. & Chang, J.H., 2004. High-resolution analysis of early diagenetic effects on magnetic minerals in post-middle-Holocene continental shelf sediments from the Korea Strait, *J. Geophys. Res.*, 109(3), 15, doi: 10.1029/2003JB002813.

Liu, J., Yang, Z., Tong, Y., Yuan, W. & Wang, B., 2010. Tectonic implications of Early–Middle Triassic palaeomagnetic results from Hexi Corridor, north China, *Geophys. J. Int.*, 182, 1216–1228.

Liu, Q., Roberts, A.P., Rohling, E.J., Zhu, R. & Sun, Y., 2008. Post-depositional remanent magnetization lock-in and the location of the Matuyama-Brunhes geomagnetic reversal boundary in marine and Chinese loess sequences, *Earth Planet. Sci. Lettrs*, 275, 102–110.

Liu, Z., Zhao, X., Wang, C., Liu, S. & Yi, H., 2003. Magnetostratigraphy of Tertiary sediments from the Hoh Xil Basin; implications for the Cenozoic tectonic history of the Tibetan Plateau, *Geophys. J. Int.*, 154, 233–252.

Lovlie, R., 1974. Post-depositional remanent magnetization in a re-deposited deep-sea sediment, *Earth Planet. Sci. Lettrs*, 21, 315–320.

Lovlie, R. & Torsvik, T., 1984. Magnetic remanence and fabric properties of laboratory-deposited hematite-bearing red sandstone, *Geophys. Res. Lettrs*, 11, 221–224.

Lowrie, W., 1990. Identification of ferromagnetic minerals in a rock by coercivity and unblocking temperature properties, *Geophys. Res. Lettrs*, 17, 159–162.

Lu, G. & McCabe, C., 1993. Magnetic fabric determined from ARM and IRM anisotropies in Paleozoic carbonates, Southern Appalachian Basin, *Geophys. Res. Lettrs*, 20, 1099–1102.

Lu, G., McCabe, C., Henry, D.J. & Schedl, A., 1994. Origin of hematite carrying a late Paleozoic remagnetization in a quartz sandstone bed from the Silurian Rose Hill Formation, Virginia, USA, *Earth Planet. Sci. Lettrs*, 126, 235–246.

Lund, S.P. & Keigwin, L., 1994. Measurement of the degree of smoothing in sediment paleomagnetic secular variation records; an example from late Quaternary deep-sea sediments of the Bermuda Rise, western North Atlantic Ocean, *Earth Planet. Sci. Lettrs*, 122, 317–330.

Lund, S.P. & Banerjee, S.K., 1985. Late Quaternary paleomagneic field secular variation from two Minnesota lakes, *J. Geophys. Res.*, 90, 803–825.

MacDonald, G.A., 1972. *Volcanoes*, Prentice-Hall, Englewood Cliffs, NJ.

Madsen, K.N., Walderhaug, H. & Torsvik, T. 2002. Erroneous fold tests as an artifact of alteration chemical remanent magnetization, *J. Geophys. Res.*, 107(11), doi: 10.1029/2001JB000805.

Maher, B.A., 1999. *Quaternary Climates, Environments and Magnetism*, Cambridge University Press, Cambridge, UK.

Maher, B.A. & Hallam, D.F., 2005. Magnetic carriers and remanence mechanisms in magnetite-poor sediments of Pleistocene age, southern North Sea margin, *J. Quat. Sci.*, 20, 79–94.

Maher, B.A., Thompson, R. & Zhou, L.-P., 1994. Spatial and temporal reconstructions of changes in the Asian palaeomonsoon: A new mineral magnetic approach, *Earth Planet. Sci. Lettrs*, 125, 462–471.

Mann, M.E. & Lees, J., 1996. Robust estimation of background noise and signal detection in climatic time series, *Clim. Change*, 33, 409–445.

Marangon, A., 2011. Stratigraphic analysis on Monte Agnello and Latemar platforms, PhD thesis, University of Padua, Italy.

March, A., 1932. Mathematische theorie der regelung nach der korngetstalt, *Z. Kristallogr.*, 82, 285–297.

Marco, S., Ron, H., McWilliams, M.O. & Stein, M., 1998. High-resolution record of geomagnetic secular variation from late Pleistocene Lake Lisan sediments (paleo Dead Sea), *Earth Planet. Sci. Lettrs*, 161, 145–160.

Mayer, H. & Appel, E., 1999. Milankovitch cyclicity and rock-magnetic signatures of palaeoclimatic change in the Early Cretaceous Biancone Formation of the Southern Alps, Italy, *Cretaceous Research*, 20, 189–214.

McCabe, C. & Elmore, R.D., 1989. The occurrence and origin of Late Paleozoic remagnetization in the sedimentary rocks of North America, *Rev. of Geophys.*, 27, 471–494.

McCabe, C. & Channell, J.E.T., 1994. Late Paleozoic remagnetization in limestones of the Craven Basin (northern England) and the rock magnetic fingerprint of remagnetized sedimentary carbonates, *J. Geophys. Res.*, 99, 4603–4612.

McCabe, C., Sassen, R. & Saffer, B., 1987. Occurrence of secondary magnetite within biodegraded oil, *Geology*, 15, 7–10.

McCabe, C., Jackson, M. & Saffer, B., 1989. Regional patterns of magnetite authigenesis in the Appalachian Basin; implications for the mechanism of late Paleozoic remagnetization, *J. Geophys. Res.*, 94, 10429–10443.

McCall, A.M. & Kodama, K.P., 2010. Inclination correction for the Moenave Formation and Wingate Sandstone: Implications for North America's apparent polar wander path and Colorado Plateau rotation. in *AGU 2010 Fall Meeting*, American Geophysical Union, San Francisco, CA.

McElhinny, M.W., 1964. Statistical significance of the fold test in paleomagnetism, *Geophys. J. Roy. Astron. Soc.*, 8, 338–340.

McElhinny, M.W. & Lock, J., 1990. IAGA global palaeomagnetic database, *Geophys. J. Int.*, 101, 763–766.

McElhinny, M.W. & McFadden, P.L., 1997. Palaeosecular variation over the past 5 Myr based on a new generalized database, *Geophys. J. Int.*, 131, 240–252.

McElhinny, M.W. & McFadden, P.L., 2000. *Paleomagnetism, Continents and Oceans*, Academic Press, San Diego.

McFadden, P.L., 1990. A new fold test for paleomagnetic studies, *Geophys. J. Int.*, 103, 163–169.

McFadden, P.L. & Jones, D.L., 1981. The fold test in paleomagnetism, *Geophys. J. Roy. Astron. Soc.*, 67, 53–58.

McFadden, P.L. & Lowes, F.J., 1981. The discrimination of mean directions drawn from Fisher distributions, *Geophys. J. Roy. Astron. Soc.*, 67, 19–33.

McFadden, P.L. & McElhinny, M.W., 1990. Classification of the reversals test in paleomagnetism, *Geophys. J. Int.*, 103, 725–729.

McIntosh, W.C., Hargraves, R.B. & West, C.L., 1985. Paleomagnetism and oxide mineralogy of Upper Triassic to Lower Jurassic red beds and basalts in the Newark Basin, *Geol. Soc. Am. Bull.*, 96, 463–480.

McNeill, D.F., 1997. Facies and early diagenetic influence on the depositional magnetization of carbonates, *Geology*, 25, 799–802.

Meijers, M.J.M., Hamers, M.F., van Hinsbergen, D.J.J., van der Meer, D.G., Ktchka, A., Langereis, C.G. & Stephenson, R.A., 2010a. New late Paleozoic paleopoles from the Donbas Foldbelt (Ukraine); implications for the Pangea A vs. B controversy, *Earth Planet. Sci. Lettrs*, 297, 18–33.

Meijers, M.J.M., Langereis, C.G., van Hinsbergen, D.J.J., Kaymakci, N., Stephenson, R.A. & Altiner, D., 2010b. Jurassic-Cretaceous low paleolatitudes from the circum-Black Sea region (Crimea and Pontides) due to true polar wander, *Earth Planet. Sci. Lettrs*, 296, 210–226.

Menke, W. & Abbott, D., 1990. *Geophysical Theory*, Columbia University Press, New York.

Merrill, R.T., McElhinny, M.W. & McFadden, P.L., 1996. *The Magnetic Field of the Earth, Paleomagnetism, the Core and the Deep Mantle*, Academic Press, San Diego.

Metcalfe, R., Rochelle, C.A., Savage, D. & Higgo, J.W., 1994. Fluid-rock interactions during continental red bed diagenesis; implications for theoretical models of mineralization in sedimentary basins. In *Geofluids: Origin, Migration and Evolution of Fluids in Sedimentary Basins*, Parnell, J. (ed.), Geological Society, London, Special Publication, 78, 301–324.

Miller, J.D. & Kent, D.V., 1986. Paleomagnetism of the Upper Devonian Catskill Formation from the southern limb of the Pennsylvania salient: Possible evidence of oroclinal rotation, *Geophys. Res. Lettrs*, 13, 1173–1176.

Mitra, R. & Tauxe, L., 2009. Full vector model for magnetization in sediments, *Earth Planet. Sci. Lettrs*, 286, 535–545.

Molina-Garza, R.S., Geissman, J.W. & Lucas, S.G., 2003. Paleomagnetism and magnetostratigraphy of the lower Glen Canyon and upper Chinle Groups, Jurassic-Triassic of northern Arizona and northeast Utah, *J. Geophys. Res.*, 108(B4), 2181.

Morrish, A.H., 1994. Canted antiferromagnetism: hematite, World Scientific Publishing Co., Singapore, 192 pp.

Mommersteeg, H.J.P.M., Loutre, M.-F., Young, R. & Hooghiemstra, H., 1995. orbital forced frequencies in the 975000 year pollen record from Tenagi Philippon (Greece), *Clim. Dyn.*, 11, 4–24.

Moskowitz, B.M., Frankel, R.B. & Bazylinski, D.A., 1993. Rock magnetic criteria for the detection of biogenic magnetite, *Earth Planet. Sci. Lettrs*, 120, 283–300.

Moskowitz, B.M., Bazylinski, D.A., Egli, R., Frankel, R.B. & Edwards, K.J., 2008. Magnetic properties of marine magnetotactic bacteria in a seasonally stratified coastal pond (Salt Pond, MA, USA), *Geophys. J. Int.*, 174, 75–92.

Mundil, R., Zuhlke, R., Bechstadt, T., Peterhansel, A., Egenhoff, S.O., Oberli, F., Meiter, M., Brack, P. & Rieber, H., 2003. Cyclicities in Triassic platform carbonates: synchronizing radio-isotopic and orbital clocks, *Terra Nova*, 15, 81–87.

Nagata, T., 1961. *Rock Magnetism*, Maruzen Co., Tokyo.

Nickelsen, R.P., 1979. Sequence of structural stages of the Alleghany Orogeny at the Bear Valley strip mine, Shamokin, Pennsylvania, *Am. J. Sci.*, 279, 225–271.

Noel, N., 1980. Surface tension phenomena in the magnetization of sediments, *Geophys. J. Roy. Astron. Soc.*, 62, 15–25.

Nurgaliev, D.K., Borisov, A.S., Heller, F., Burov, B.V., Jasonov, P.G., Khasanov, D.I. & Ibragimov, S.Z., 1996. Geomagnetic secular variation through the last 3500 years as recorded by Lake Aslikul sediments from eastern Europe (Russia), *Geophys. Res. Lettrs*, 23, 375–378.

Oda, H. & Torii, M., 2004. Sea-level change and remagnetization of continental shelf sediments off New Jersey (ODP Leg 174A): Magnetite and greigite diagenesis, *Geophys. J. Int.*, 156, 443–458.

Ojha, T.P., Butler, R.F., Quade, J., DeCelles, P.G., Richards, D. & Upreti, B.N., 2000. Magnetic polarity stratigraphy of the Neogene Siwalik Group at Khutia Khola, far western Nepal, *Geol. Soc. Am. Bull.*, 112, 424–434.

Oldfield, F., 1990. Magnetic measurements of Recent sediments from Big Moose Lake, Adirondack Mountains, NY USA, *J. Paleolimnology*, 4, 93–101.

Oldfield, F., 1994. Toward the discrimination of fine grained ferrimagnetics by magnetic measurements in lake and nearshore marine sediments, *J. Geophys. Res.*, 99, 9045–9050.

Oldfield, F., Hunt, A., Jones, D.H., Chester, R., Deargin, J.A., Olsson, L. & Prospero, J.M., 1985. Magnetic differentiation of atmospheric dusts, *Nature*, 317, 516–518.

Oliver, J., 1986. Fluids expelled tectonically from orogenic belts: Their role in hydrocarbon migration and other geologic phenomena, *Geology*, 14, 99–102.

Olsen, P.E., 1997. Stratigraphic record of the early Mesozoic breakup of Pangea in the Laurasia-Gondwana rift system, *Ann. Rev. of Earth Planet. Sci.*, 25, 337–401.

Ong, P.F., van der Pluijm, B. & Van der Voo, R., 2007. Early rotation and late folding in the Pennsylvania salient (U.S. Appalachians): Evidence from calcite-twinning analysis of Paleozoic carbonates, *Geol. Soc. Am. Bull.*, 119, 796–804.

Opdyke, N.D. & Henry, K.W., 1969. A test of the dipole hypothesis, *Earth Planet. Sci. Lettrs*, 6, 139–151.

Otofuji, Y. & Sasajima, S., 1981. A magnetization process of sediments: laboratory experiments on post-depositional remanent magnetization, *Geophys. J. Roy. Astron. Soc.*, 66, 241–259.

Parrish, J.T., 1998. *Interpreting Pre-Quaternary Climate from the Geologic Record*, Columbia University Press, New York, NY.

Patzelt, A., Li, H., Wang, J. & Appel, E., 1996. Palaeomagnetism of Cretaceous to Tertiary sediments from southern Tibet: evidence for the extent of the northern margin of India prior to the collision with Eurasia, *Tectonophysics*, 259, 259–284.

Payne, M.A. & Veosub, K.L., 1982. The acquisition of postdepositional detrital remanent magnetism in a variety of natural sediments, *Geophys. J. Roy. Astron. Soc.*, 68, 625–642.

Peng, L. & King, J.W., 1992. A Late Quaternary geomagnetic secular variation record from Lake Waiau, Hawaii, and the question of the Pacific nondipole low, *J. Geophys. Res.*, 97, 4407–4424.

Peters, C. & Dekkers, M.J., 2003. Selected room temperature magnetic parameters as a function of mineralogy, concentration, and grain size, *Phys. Chem. Earth*, 28, 659–667.

Petersen, N., von Dobeneck, T. & Vali, H., 1986. Fossil bacterial magnetite in deep-sea sediments from the South Atlantic Ocean, *Nature*, 320, 611–615.

Piper, J.D.A., Mesci, L.V., Gursoy, H., Tatar, O. & Davies, C.J., 2007. Palaeomagnetic and rock magnetic properties of travertine; its potential as a recorder of geomagnetic pal- aeosecular variation, environmental change and earth- quake activity in the Sicak Cermik geothermal field, Turkey, *Phys Earth Planet. Int.*, 161, 50–73.

Preto, N., Hinnov, L.A., Hardie, L.A. & De Zache, V., 2001. Middle Triassic orbital signature recorded in the shallow- marine Latemar carbonate build-up (Dolomites, Italy), *Geology*, 29, 1123–1126.

Preto, N., Spotl, C., Paola, G., Riva, P.M. & Manfrin, S., 2007. Aragonite dissolution, sedimentation rates and carbon iso- topes in deep-water hemipelagites (Livinallongo Formation, Middle Triassic, northern Italy), Reply, *Sedimentary Geology*, 194, 101–110.

Pullaiah, G.E., Irving, E., Buchan, K.L. & Dunlop, D.J., 1975. Magnetization changes caused by burial and uplift, *Earth Planet. Sci. Lettrs*, 28, 133–143.

Quidelleur, X. & Courtillot, V., 1996. On low-degree spherical harmonic models of paleosecular variation, *Phys Earth Planet. Int.*, 95, 55–77.

Ramsay, J.G., 1967. *Folding and Fracturing of Rocks*, McGraw- Hill, New York.

Ramsay, J.G. & Huber, M.I., 1983. *The Techniques of Modern Structural Geology, Vol. 1, Strain Analysis*, Academic, Orlando, FL.

Rapalini, A.E., 2006. New late Proterozoic paleomagnetic pole for the Rio de la Plata Craton; implications for Gond- wana, *Precamb. Res.*, 147, 223–233.

Rapalini, A.E., Fazzito, S. & Orue, D., 2006. A new late Permian paleomagnetic pole for stable South America; the Independencia Group, eastern Paraguay, *Earth, Planets, Space*, 58, 1247–1253.

Rees, A.I., 1961. The effect of water currents on the magnetic remanence and anisotropy of susceptibility of some sedi- ments, *Geophys. J. London*, 8, 235–251.

Renne, P.R. & Onstott, T.C., 1988. Laser-selective demagneti- zation; a new technique in paleomagnetism and rock mag- netism, *Science*, 242, 1152–1155.

Reynolds, R.L., Fishman, N.S., Hudson, M.R., Karachewski, J.A. & Goldhaber, M.B., 1985. Magnetic minerals and hydrocarbon migration: Evidence from Cement (Okla- homa), north slope (Alaska), and the Wyoming-Idaho- Utah thrust belt, *Eos Trans. AGU*, 66, 867.

Reynolds, R.L., Tuttle, M.L., Rice, C.A., Fishman, N.S., Kara- chewski, J.A. & Sherman, D.M., 1994. Magnetization and geochemistry of greigite-bearing Cretaceous strata, North Slope basin, Alaska, *Am. J. Sci.*, 294, 485–528.

Rial, J.A., Pielke, R.A., Beniston, M., Claussen, M., Canadell, J., Cox, P., Held, H., de Noblet-Ducoudre, N., Prinn, R., Rey- nolds, J.F. & Salas, J., 2004. Nonlinearities, feedbacks and critical thresholds within the Earth's climate system, *Clim. Change*, 65, 11–38.

Richter, C., Roberts, A.P., Stoner, J.S., Benning, L.D. & Chi, C.T., 1998. Magnetostratigraphy of Pliocene-Pleistocene sedi- ments from the eastern Mediterranean Sea, *Proc. Ocean Drill. Prog, Scientific Results*, 160, 61–73.

Roberts, A.P. & Turner, G.M., 1993. Diagenetic formation of ferrimagnetic iron sulphide minerals in rapidly deposited marine sediments, South Island, New Zealand, *Earth Planet. Sci. Lettrs*, 115, 257–273.

Roberts, A.P. & Weaver, R., 2005. Multiple mechanisms of remagnetization involving sedimentary greigite (Fe3S4), *Earth Planet. Sci. Lettrs*, 231, 263–277.

Roberts, A.P., Jiang, W.T., Florindo, F., Horng, C.-S. & Laj, C., 2005. Assessing the timing of greigite formation and the reliability of the upper Olduvai polarity transition record from the Crostolo River, Italy, *Geophys. Res. Lettrs*, 32, doi: 10.1029/2004GL022137.

Roberts, A.P., Florindo, F., Villa, G., Chang, L., Jovane, L., Bohaty, S.M., Lorrasoana, J.C., Heslop, D. and Fitz Gerald, J.D., 2011. Magnetotactic bacterial abundance in pelagic marine environments is limated by organic carbon flux and availability of dissolved iron, *Earth Planet. Sci. Lettrs*, 310, 441–452.

Rochette, P. & Vandamme, D., 2001. Pangea B; an artifact of incorrect paleomagnetic assumptions? *Annali di Geofisica*, 44, 649–658.

Rochette, P., Tamrat, E., Feraud, G., Pik, R., Courtillot, V., Ketefo, E., Coulon, C., Hoffman, C., Vandamme, D. & Yirgu, G., 1988. Magnetostratigraphy and timing of the Oligocene Ethiopian traps, *Earth Planet. Sci. Lettrs*, 164, 497–510.

Rowan, C.J. & Roberts, A.P., 2006. Magnetite dissolution, diachronous greigite formation, and secondary magnet- izations from pyrite oxidation: Unravelling complex mag- netizations in Neogene marine sediments from New Zealand, *Earth Planet. Sci. Lettrs*, 241, 119–137.

Rowan, C.J., Roberts, A.P. & Broadbent, T., 2009. Reductive diagenesis, magnetite dissolution, greigite growth and paleomagnetic smoothing in marine sediments: A new view, *Earth Planet. Sci. Lettrs*, 277, 223–235.

Sadler, P.M., 1981. Sediment accumulation rates and the completeness of stratigraphic sections, *J. Geology*, 89, 569–584.

Sager, W.W. & Singleton, S.W., 1989. Paleomagnetic inclina- tion errors in sediments cored from a Gulf of Mexico salt dome; an indicator of ancient sea level? *Geology*, 17, 739–742.

Sagnotti, L., Roberts, A.P., Weaver, R., Verosub, K.L., Florindo, F., Pike, C.R., Clayton, T. & Wilson, G.S., 2005. Apparent magnetic polarity reversals due to remagnetization result- ing from late diagenetic growth of greigite from siderite, *Geophys. J. Int.*, 160, 89–100.

Sagnotti, L., Cascella, A., Ciaranfi, N., Macri, P., Maiorano, P., Marino, M. & Taddeucci, J., 2010. Rock magnetism and paleomagnetic of the Montalbano Jonico section (Italy): evidence for late diagenetic growth of greigite and implica- tions for magnetostratigraphy, *Geophys. J. Int.*, 180, 1049–1066.

Shcherbakov, V. & Shcherbakova, V., 1983. On the theory of depositional remanent magnetization in sedimentary rocks, *Geophys. Surv.*, 5, 369–380.

Schmidt, P.W., Williams, G.E. & McWilliams, M.O., 2009. Palaeomagnetism and magnetic anisotropy of late Neoproterozoic strata, South Australia: Implications for the palaeolatitude of late Cryogenian glaciation, cap carbonate and the Ediacaran System, *Precamb. Res.*, 174, 35–52.

Schulz, H.D. & Zabel, M., 2006. *Marine Geochemistry*, Springer-Verlag, Berlin.

Schwehr, K., Tauxe, L., Driscoll, N. & Lee, H., 2006. Detecting compaction disequilibrium with anisotropy of magnetic susceptibility, *Geochem. Geophys. Geosyst.*, 7, Q11002.

Scotese, C.R., Van der Voo, R. & McCabe, C., 1982. Paleomagnetism of the Upper Silurian and Lower Devonian carbonates of New York State; evidence for secondary magnetizations residing in magnetite, *Phys Earth Planet. Int.*, 30, 385–395.

Shaw, J., 2007. Microwave paleomagnetic technique. In *Encyclopedia of Earth Sciences*, Gubbins, D. & Herrero-Bervera, E. (eds) Springer, Dordrecht, Netherlands, pp. 694–695.

Sheldon, N.D., 2005. Do red beds indicate paleoclimatic conditions? A Permian case study, *Paleogeog., Paleoclimat. Paleoecol.*, 228, 305–319.

Smethurst, M.A. & McEnroe, S.A., 2003. The palaeolatitude controversy in the Silurian of Newfoundland resolved; new palaeomagnetic results from the Central Mobile Belt, *Tectonophysics*, 362, 83–104.

Snowball, I., Zillen, L. & Sandgren, P., 2002. Bacterial magnetite in Swedish varved lake-sediments: a potential bio-marker of environmental change, *Quatern. Int.*, 88, 13–19.

Spahn, Z., 2011. Toward resolving the Latemar Controversy: A Magnetostratigraphy of the Latemar correlated sequence at Rio Sacuz, Masters thesis, Lehigh University, Bethlehem, PA.

Spahn, Z. & Kodama, K.P., 2011. Resolving the Latemar Controversy: a new magnetostratigraphy at Rio Sacuz, Abstract GP53A-08, 2011 Fall Meeting, AGU, San Francisco, CA 5–9 December 2011.

Stacey, F.D., 1972. On the role of Brownian motion in the control of detrital remanent magnetization of sediments, *Pure Appl. Geophys.*, 98, 139–145.

Stamatakos, J. & Kodama, K.P., 1991a. Flexural flow folding and the paleomagnetic fold test: An example of strain reorientation of remanence in the Mauch Chunk formation, *Tectonics*, 10, 807–819.

Stamatakos, J. & Kodama, K.P., 1991b. The effects of grain-scale deformation on the Bloomsburg formaiton pole, *J. Geophys. Res.*, 96, 17919–17933.

Stamatakos, J. & Hirt, A.M., 1994. Paleomagnetic considerations of the development of the Pennsylvania salient in the central Appalachians, *Tectonophysics*, 231, 237–255.

Stamatakos, J., Van der Voo, R., van der Pluijm, B., Potts, S., Torsvik, T., Hodych, J.P. & Buchan, K.L., 1994. Comment and reply on Early Silurian palaeolatitude of the Springdale Group redbeds of central Newfoundland; a palaeomagnetic determination with a remanence anisotropy test for inclination error (modified), *Geophys. J. Int.*, 119, 1009–1015.

Stamatakos, J., Hirt, A.M. & Lowrie, W., 1996. The age and timing of folding in the central Appalachians from paleomagnetic results, *Geol. Soc. Am. Bull.*, 108, 815–829.

Stead, R.J. & Kodama, K.P., 1984. Paleomagnetism of the Cambrian rocks of the Great Valley of east-central PA: Fold test constraints on the age of magnetization. In *Plate Reconstruction from Paleozoic paleomagnetism*, Van der Voo, R., Scotese, C. R. & Bonhommet, N. (eds), American Geophysical Union, Washington, DC, pp. 120–130.

Steiner, M.B., 1983. Detrital remanent magnetization in hematite, *J. Geophys. Res.*, 88, 6523–6539.

Stober, J.C. & Thompson, R., 1979. An investigation into the source of magnetic minerals in some Finnish lake sediments, *Earth Planet. Sci. Lettrs*, 45, 464–474.

Stockhausen, H., 1998. Geomagnetic paleosecular variation (0–13000 yr BP) as recorded in sediments from three maar lakes from the West Eifel (Germany), *Geophys. J. Int.*, 135, 898–910.

Stokking, L.B. & Tauxe, L., 1990a. Multicomponent magnetization in synthetic hematite, *Phys Earth Planet. Int.*, 65, 109–124.

Stokking, L.B. & Tauxe, L., 1990b. Properties of chemical remanence in synthetic hematite; testing theoretical predictions, *J. Geophys. Res.*, 95, 12, 639–12, 652.

Suganuma, Y., Yokoyama, Y., Yamazaki, T., Kawamura, K., Horng, C.-S. & Matsuzaki, H., 2010. [10]Be evidence for delayed acquisition of remanent magnetization in marine sediments: Implication for a new age for the Matuyama-Brunhes boundary, *Earth Planet. Sci. Lettrs*, 296, 443–450.

Suganuma, Y., Okuno, J., Heslop, D., Roberts, A.P., Yamazaki, T. & Yokoyama, Y., 2011. Post-depositional remanent magnetization lock-in for marine sediments deduced from [10]Be and paleomagnetic records through the Matuyama-Brunhes boundary, *Earth Planet. Sci. Lettrs*, 311, 39–52.

Suk, D., Van der Voo, R. & Peacor, D.R., 1990. Scanning and transmission electron microscope observations of magnetite and other iron phases in Ordovician carbonates from East Tennessee, *J. Geophys. Res.*, 95, 12327–12336.

Sun, W.-W. & Kodama, K.P., 1992. Magnetic anisotropy, scanning electron microscopy, and x-ray pole figure goniometry study of inclination shallowing in a compacting clay-rich sediment, *J. Geophys. Res.*, 97, 19599–19615.

Sun, W.-W., Jackson, M. & Craddock, J.P., 1993. Relationship between remagnetization, magnetic fabric and deformation in Paleozoic carbonates, *Tectonophysics*, 221, 361–366.

Sun, Z., Yang, Z., Pei, J., Yang, T. & Wang, X., 2006. New Early Cretaceous paleomagnetic data from volcanic and red beds of the eastern Qaidam Block and its implications for tectonics of Central Asia, *Earth Planet. Sci. Lettrs*, 243, 268–281.

Tamaki, M. & Itoh, Y., 2008. Tectonic implications of paleomagnetic data from Upper Cretaceous sediments in the

Oyubari area, central Hokkaido, Japan, *Island Arc*, 17, 270–284.

Tamaki, M., Oshimbe, S. & Itoh, Y., 2008. A large latitudinal displacement of a part of Cretaceous fore-arc basin in Hokkaido, Japan; paleomagnetism of the Yezo Supergroup in the Urakawa area, *J. Geol. Soc. Japan*, 114, 207–217.

Tan, X. & Kodama, K.P., 1998. Compaction-corrected inclinations from southern California Cretaceous marine sedimentary rocks indicate no paleolatitudinal offset for the Peninsular Ranges terrane, *J. Geophys. Res.*, 103, 27169–27192.

Tan, X. & Kodama, K.P., 2002. Magnetic anisotropy and paleomagnetic inclination shallowing in red beds: Evidence from the Mississippian Mauch Chunk Formation, Pennsylvania, *J. Geophys. Res.*, 107, doi: 10/1029/2001JB001636.

Tan, X. & Kodama, K.P., 2003. An analytical solution for correcting palaeomagnetic inclination error, *Geophys. J. Int.*, 152, 228–236.

Tan, X., Kodama, K.P. & Fang, D., 2002. Laboratory depositional and compaction-caused inclination errors carried by haematite and their implications in identifying inclination error of natural remanence in red beds, *Geophys. J. Int.*, 151, 475–486.

Tan, X., Kodama, K.P., Chen, H., Fang, D., Sun, D. & Li, Y., 2003. Paleomagnetism and magnetic anisotropy of Cretaceous red beds from the Tarim basin, northwest China: Evidence for a rock magnetic cause of anomalously shallow paleomagnetic inclinations from central Asia, *J. Geophys. Res.*, 108, doi: 10.1029/2001JB00160.

Tan, X., Kodama, K.P., Gilder, S. & Courtillot, V., 2007. Rock magnetic evidence for inclination shallowing in the Passaic Formation red beds from the Newark basin and a systematic bias of the Late Triassic apparent polar wander path for North America, *Earth Planet. Sci. Lettrs*, 254, 345–357.

Tan, X., Gilder, S., Kodama, K.P., Jiang, W., Han, Y., Zhang, H., Xu, H. & Zhou, D., 2010. New paleomagnetic results from the Lhasa block: Revised estimation of latitudinal shortening across Tibet and implications for dating the India-Asia collision, *Earth Planet. Sci. Lettrs*, 293, 396–404.

Tarduno, J.A., 1994. Temporal trends of magnetic dissolution in the pelagic realm: Gauging paleoproductivity? *Earth Planet. Sci. Lettrs*, 123, 39–48.

Tarling, D.H., 1999. Palaeoamgnetism and diagenesis in sediments. In *Palaeomagnetism and Diagenesis in Sediments*, Tarling, D.H. & Turner, P. (eds), Geological Society, London, Special Publication 151, 1–8.

Tauxe, L., 2005. Inclination flattening and the geocentric axial dipole hypothesis, *Earth Planet. Sci. Lettrs*, 233, 247–261.

Tauxe, L., 2010. *Essentials of Paleomagnetism*, University of California Press, Berkeley.

Tauxe, L. & Opdyke, N.D., 1982. A time framework based on magnetostratigraphy for the Siwalik sediments of the Khaur area, northern Pakistan, *Paleogeog., Paleoclimat. Paleoecol.*, 37, 43–61.

Tauxe, L. & Kent, D.V., 1984. Properties of a detrital remanence carried by haematite from study of modern river deposits and laboratory redeposition experiments, *Geophys. J. Roy. Astron. Soc.*, 76, 543–561.

Tauxe, L. & Watson, G.S., 1994. The fold test: an eigen analysis approach, *Earth Planet. Sci. Lettrs*, 122, 331–341.

Tauxe, L. & Hartl, P., 1997. 11 milion years of Oligocene geomagnetic field behaviour, *Geophys. J. Int.*, 128, 217–229.

Tauxe, L. & Kent, D.V., 2004. A simplified statistical model for the geomagnetic field and the detection of shallow bias in paleomgnetic inclinations; was the ancient magnetic field dipolar? In *Geophysical Monograph*, Channell, J.E.T., Kent, D.V., Lowrie, W. & Meert, J.G. (eds), American Geophysical Union, Washington, DC, 101–115.

Tauxe, L. & Kodama, K.P., 2009. Paleosecular variation models for ancient times; clues from Keweenawan lava flows, *Phys Earth Planet. Int.*, 177, 31–45.

Tauxe, L., Herbert, T., Shackleton, N.J. & Kok, Y.S., 1996. Astronomical calibration of the Matuyama-Brunhes boundary: Consequences for magnetic remanence acquisition in marine carbonates and the Asian loess sequences *Earth Planet. Sci. Lettrs*, 140, 133–146.

Tauxe, L., Bertram, H.N. & Seberino, C., 2002. Physical interpretation of hysteresis loops: Micromagnetic modeling of fine particle magnetite, *Geochem. Geophys. Geosyst.*, 3, doi: 10.1029/2001GC000241.

Tauxe, L., Steindorf, J.L. & Harris, A., 2006. Depositional remanent magnetization; toward an improved theoretical and experimental foundation, *Earth Planet. Sci. Lettrs*, 244, 515–529.

Tauxe, L., Kodama, K.P. & Kent, D.V., 2008. Testing corrections for paleomagnetic inclination error in sedimentary rocks; a comparative approach, *Phys Earth Planet. Int.*, 169, 152–165.

Taylor, G.K., Tucker, C., Twitchett, R.J., Kearsey, T., Benton, M.J., Newell, A.J., Surkov, M.V. & Tverdokhlebov, V.P., 2009. Magnetostratigraphy of Permian/Triassic boundary sequences in the Cis-Urals: No evidence for a major temporal hiatus, *Earth Planet. Sci. Lettrs*, 281, 36–47.

Thompson, R. & Oldfield, F., 1986. *Environmental Magnetism*, Allen & Unwin, Boston, 227 pp.

Thompson, R. & Cameron, T.D.J., 1995. Palaeomagnetic study of Cenozoic sediments in North Sea boreholes: An example of a magnetostratigraphic conundrum in a hydrocarbon-producing area. In *Palaeomagnetic Applications in Hydrocarbon Exploration*, Turner, P. & Turner, A. (eds), Geological Society, London, Special Publication, 98, pp. 223–236.

Thompson, L.G., Yao, T., Davis, M.E., Henderson, K.A. & Mosley-Thomson, E., 1997. Tropical climate instability: the last glacial cycle from a Qinhai-Tibetan ice core, *Science*, 276, 1821–1825.

Thomson, D.J., 1982. Spectrum estimation and harmonic analysis, In *Proceedings of the IEEE*, 70, 1055–1096.

Till, J.L., Jackson, M. & Moskowitz, B.M., 2010. Remanence stability and magnetic fabric development in synthetic shear zones deformed at 500°C, *Geochem. Geophys. Geosyst.*, 11, doi: 10.1029/2010GC003320.

Tohver, E., Weil, A.B., Solum, J.G. & Hall, C.M., 2008. Direct dating of carbonate remagnetization by ^{40}Ar/^{39}Ar analysis of the smectite-illite transformation, *Earth and Planet. Sci. Lettrs*, 274, 524–530.

Torcq, F., Besse, J., Vaslet, D., Marcoux, J., Ricou, L.E., Halawani, M. & Basahel, M., 1997. Paleomagnetic results from Saudi Arabia and the Permo-Triassic Pangea configuration, *Earth Planet. Sci. Lettrs*, 148, 553–567.

Torsvik, T. & Van der Voo, R., 2002. Refining Gondwana and Pangea palaeogeography; estimates of Phanerozoic non-dipole (octupole) fields, *Geophys. J. Int.*, 151, 771–794.

Torsvik, T.H., Muller, R.D., Van der Voo, R., Steinberger, B. & Gaina, C., 2008. Global plate motion frames: toward a unified model, *Rev. of Geophys.*, 46, RG3004.

Tric, E., Laj, C., Jehanno, C., Valet, J.-P., Kissel, C., Mazaud, A. & Iaccarino, S., 1991. High-resolution record of the Upper Olduvai transition from Po Valley (Italy) sediments: Support for dipolar transition geometry? *Phys Earth Planet. Int.*, 65, 319–338.

Tucker, P., 1980. A grain mobility model of post-depositional realignment, *Geophys. J. Roy. Astron. Soc.*, 63, 149–163.

Turner, P., 1980. *Continental Red Beds*, Elsevier, Amsterdam.

Valet, J.-P., Meynadier, L. & Guyodo, Y., 2005. Geomagnetic dipole strength and reversal rate over the past two million years, *Nature*, 435, 802–805.

van der Pluijm, B., 1987. Grain-scale deformation and the fold test-evaluation of syn-folding remagnetization, *Geophys. Res. Lettrs*, 14, 155–157.

Van der Voo, R., 1990. Phanerozoic paleomagnetic poles from Europe and North America and comparisons with contiental reconstructions, *Rev. of Geophys.*, 28, 167–206.

Van der Voo, R., 1993. *Paleomagnetism of the Atlantic, Tethys, and Iapetus Oceans*, Cambridge University Press, Cambridge.

Van der Voo, R., & Torsvik, T., 2012. Remagnetization history. In *Remagnetization and Chemical Alteration of Sedimentary Rocks*, Elmore R.D., Muxworthy, A.R., Aldana, M.M. & Mena, M. (eds), Geological Society, London, Special Publications, 371.

Van Houten, F.B., 1973. Origin of red beds a review; 1961–1972, *Ann. Rev. of Earth Planet. Sci.*, 1, 39–61.

Vasiliev, K., Dekkers, M.J., Krijgsman, W., Franke, C., Langereis, C.G. & Mullender, T.A.T., 2007. Early diagenetic greigite as a recorder of the palaeomagnetic signal in Miocene-Pliocene sedimentary rocks of the Carpathian foredeep (Romania), *Geophys. J. Int.*, 171, 613–629.

Vaughn, J., Kodama, K.P. & Smith, D.P., 2005. Correction of inclination shallowing and its tectonic implications: The Cretaceous Perforada Formation, Baja California, *Earth Planet. Sci. Lettrs*, 232, 71–82.

Verosub, K.L., 1977. Depositional and postdepositional processes in the magnetization of sediments, *Rev. of Geophys.*, 15, 129–143.

Verosub, K.L. & Roberts, A.A., 1995. Environmental magnetism; past, present, and future, *J. Geophys. Res.*, 100, 2175–2192.

Walderhaug, H., 1992. Directional properties of alteration CRM in basic igneous rocks, *Geophys. J. Int.*, 111, 747–756.

Walker, T.R., 1967. Color of Recent sediments in tropical Mexico; a contribution to the origin of red beds, *Geol. Soc. Am. Bull.*, 78, 917–919.

Walker, T.R., 1974. Formation of red beds in moist tropical climates: a hypothesis, *Geol. Soc. Am. Bull.*, 85, 633–638.

Walker, T.R., Waugh, B. & Grone, A.J., 1978. Diagenesis in first-cycle desert alluvium of Cenozoic age, southwestern United States and northwestern Mexico, *Geol. Soc. Am. Bull.*, 89, 19–32.

Walker, T.R., Larson, E.E. & Hoblitt, R.P., 1981. The nature and origin of hematite in the Moenkopi Formation (Triassic), Colorado Plateau: A contribution to the origin of magnetism in red beds, *J. Geophys. Res.*, 86, 317–333.

Wang, B. & Yang, Z., 2007. Late Cretaceous paleomagnetic results from southeastern China, and their geological implication, *Earth Planet. Sci. Lettrs*, 258, 315–333.

Ward, P.D., Hurtado, M., Kirschvink, J.L. & Verosub, K.L., 1997. Measurements on the Cretaceous paleolatitude of Vancourver Island: Consistent with the Baja-British Columbia hypothesis, *Science*, 277, 1642–1645.

Watson, G.S., 1956. A test of randomness of directions, *Mon. Not. R. Astron. Soc. Geophys. Suppl.*, 7, 160–161.

Weaver, R., Roberts, A.P. & Barker, A.J., 2002. A late diagenetic (synfolding) magnetization carried by pyrrhotite: implications for paleomagnetic studies from magnetic iron sulphide-bearing sediments, *Earth Planet. Sci. Lettrs*, 200, 371–386.

Weedon, G., 2003. *Time-series Analysis and Cyclostratigraphy: Examining Stratigraphic Records of Environmental Cycles*, Cambridge University Press, Cambridge, UK.

Weil, A.B. & Sussman, A.J., 2004. Classifying curved orogens based on timing relationships between structural development and vertical-axis rotations, *Spec. Pap. Geol. Soc. Amer.*, 383, 1–15.

Weil, A.B., Van der Voo, R. & van der Pluijm, B., 2001. New paleomagnetic data from the southern Cantabria-Asturias Arc, northern Spain: Implications for true oroclinal rotation and the final amalgamation of Pangea, *Geology*, 29, 991–994.

Weil, A.B., Yonkee, A. & Sussman, A.J., 2010. Reconstructing the kinematic evolution of curved mountain belts: A paleomagnetic study of Triassic red beds from the Wyoming salinet, Sevier thrust belt, USA, *Geol. Soc. Am. Bull.*, 122, 3–23.

Whidden, K.J., Lund, S.P. & Bottjer, D.J., 1998. Paleomagnetic evidence that the central block of Salinia (California) is not a far-traveled terrane, *Tectonics*, 17, 329–343.

Williams, D.F., Peck, J., Karabanov, E.G., Prokopenko, A.A. & Kravchinsky, V., 1997. Lake Baikal record of continental climate response to orbital forcing during the past 5 million years, *Science*, 278, 1114–1117.

Woods, S.D., Elmore, R.D. & Engel, M.H., 2002. Paleomagnetic dating of the smectite-to-illite conversion; testing the hypothesis in Jurassic sedimentary rocks, Skye, Scotland, *J. Geophys. Res.*, 107, doi: 10.1029/2000JB000053.

Yan, M., Van der Voo, R., Tauxe, L., Fang, X. & Pares, J.M., 2005. Shallow bias in Neogene palaeomagnetic directions from the Guide Basin, NE Tibet, caused by inclination error, *Geophys. J. Int.*, 163, 944–948.

Yonkee, A. & Weil, A.B., 2010. Reconstructing the kinematic evolution of curved mountain belts; internal strain patterns in the Wyoming Salient, Sevier thrust belt, U.S.A., *Geol. Soc. Am. Bull.*, 122, 24–49.

Zhang, W., Xing, Y., Yu, L., Feng, H. & Lu, M., 2008. Distinguishing sediments from the Yangtze and Yellow Rivers, China; a mineral magnetic approach, *The Holocene*, 18, 1139–1145.

Zhao, C., Yu, Z., Ito, E., Kodama, K.P. & Chen, F., 2010. Holocene millenial-scale climate variations documented by multiple lake-level proxies in sediment cores from Hurleg Lake, Northwest China, *J. Paleolimnology*, 44, 995–1008.

Zhou, K., Zhang, W-Y., Yu-Zhang, K., Pan, H-M., Zhang, S-D., Zhang, W-J., Yue, H-D., Li, Y., Xiao, T. & Wu, L-F., 2011. A novel genus of multicellular magnetotactic prokaryotes from the Yellow Sea, *Env. Microbiol.*, doi: 10.1111/j.1462–2920.2011.02590.x.

Zijderveld, J.D.A., 1967. AC demagnetization of rocks: Analysis of results. In *Methods in Paleomagnetism*, Collinson, D.W., Creer, K.M. & Runcorn, S.K. (eds), Elsevier, Amsterdam, pp. 254–286.

Zwing, A., Clauer, N., Liewig, N. & Bachtadse, V., 2009. Identification of remagnetization processes in Paleozoic sedimentary rocks of the northeast Rhenish Massif in Germany by K-Ar dating and REE tracing of authigenic illite and Fe oxides, *J. Geophys. Res.*, 114, doi: 10.1029/2008JB006137.

Index

Note: Page references in *italics* refer to Figures; those in **bold** refer to Tables

Paleomagnetism of Sedimentary Rocks: Process and Interpretation, First Edition. Kenneth P. Kodama.
© 2012 Kenneth P. Kodama. Published 2012 by Blackwell Publishing Ltd.